Landform and Landscape in Africa

Landform and Landscape in Africa

J. M. Pritchard

Edward Arnold

First published 1979
by Edward Arnold (Publishers) Ltd
41 Bedford Square, London WC1B 3DQ

British Library Cataloguing in Publication Data
Pritchard, John Malcolm
Landform and landscape in Africa.
1. Landforms—Africa
I. Title
551.4'096   GB439

ISBN 0 7131 0204 7

Typeset by Reproduction Drawings Ltd, Sutton, Surrey.
Printed in Great Britain by Whitstable Litho Ltd,
Whitstable, Kent

Acknowledgements

The Author and Publishers wish to thank the following for
permission to use the figures listed:

AAA photo, Paris, 63, 206, 248 (Andre Picon); Aerofilms, 101,
102, 243; African Development, 103; British Steel Corporation,
247; Prof Karl Butzer & RIM Campbell, 98; J. Allan Cash, 32,
82, 233; Duggan-Cronin Collection of the McGregor Museum,
140; East-African Railways Corp. 202; Peter Fraenkel, 83, 210,
211, 228; A.T. Grove, 191, 207; Robert Harding Associates,
199; Prof R.J. Harrison Church, 34; M.I. Hassan, 104; Hoa-Qui,
222; Ministry of Information and Broadcasting, Nairobi, Kenya,
85, 174; Keystone Press Agency, 92, 231; Longman Group Ltd,
127, 130, 131; Terence J. McNally, 158; W.T.W. Morgan, 38,
40, 200; NASA, 177; Natural Science Photos, 177; Popperfoto,
22, 134, 141, 152, 194; Radio Times Hulton Picture Library,
162; Dept. of Information, South Africa, 48, 65, 90, 97; South
African Tourist Corporation, 54; Michael F. Thomas, 21, 27 and
35 from *Tropical Geomorphology: A Study in Weathering and
Landform Development in Warm Climates*, Macmillan Press,
1974; M.B. Thorp, 60; Director of Overseas Survey and Ministry
of Defence (Air Force Dept.) Crown Copyright, and Commis-
sioner of Lands and Surveys, Kampala, Uganda, 171; Dr John
B. Whittow, 145; Late Dr Ian Wilson, 217; Dr A. Young, 67;
Surveyor General's Office, Rhodesia Zimbabwe, 160; Rhodesia
Zimbabwe National Tourist Board, 106, 190;

Cover photo: Alan Hutchinson

Several of the diagrams in this book are based on the work of
others. The author would like to acknowledge his debt to the
following authors and artists:

T.N. Clifford (1); R.S. Dietz and J.C. Holden (4a and 4b);
J.F. Vine (5); A.G. Smith (6, 7 and 8); Archd. Pratt (12a) and
G. Airy (12b) — based on drawings by Bowie and Longwell;
H. Besairie (17b); M. F. Thomas (25, 35, 44); B. P. Ruxton and
L. Berry (26, 31); L.K. Jeje (37); A. Faniran and C. High (42);
D. Balazs (49); J. Nicod (50); A.S. (51); A.L. Zeitsman (53);
L.C. King (62, 118); A. Holmes (66, 101); J.C. Pugh (70);
P.P. Howell (Nile diag. in 73); N.P. Iloeje (89, 150, 151);
J.D. Clark (110); E.O. Teale and H.B.S. Cooke (114);
J.C. Doornkamp and P.H. Temple (115); E.H. Spence (123);
de Heinzelin (126); C. Downie (133); J.B. Whittow (135);
A.L.du Toit (137, 155, 188); G.W.A. Sparrow (142);
J.E. Dewey and J.M. Bird (153); C.M. Schwellnus (172);
G. Bond (173); J.H. Wellington (176, 239, 242); Report 42,
Geol. Svy of Kenya (178); E. Krenkel (182); A. Macgregor (184);
G.J.H. McCall (186); A.A. Reading (197); F. Klute (200);
B.C. King (201); A. Meigs (203); R.U. Cooke and A. Warren
(212, 214); A.T. Grove (216); R.J. Harrison Church (238);
S. Gregory (240); J. Richard-Molard (241); H. Merensky and
P.A. Wagner (243).

Contents

features; dune forms and patterns; water action in arid
regions; desert landforms caused by running water.

Preface

Landform and Landscape in Africa has been written especially
for those who are studying geomorphology at an advanced level
in schools and colleges in Africa. The book may also provide
useful background material for students living outside Africa
whose course also includes a study of the continent.

My chief purpose in writing this book is to create a greater
awareness and understanding among students living in Africa
of the processes which are moulding the landforms and landscapes of their own continent.

J. M. P.

GEOLOGICAL TABLE

Millions of years ago	Era	Periods	Systems	Continental Systems		Faulting (F), Vulcanism (V) Glaciation and Erosion Surfaces
0				Northern Africa	Southern Africa	
		Quaternary	Pleistocene	Pluvials and	Interpluvials	Glacials and interglacials. Congo erosion cycle begins.
	CAINOZOIC	Tertiary	Pliocene Miocene Oligocene Eocene	Continental Terminal	Kalahari	F V / F V / Alpine mountain building F V / African erosion begins F V
70 100		Cretaceous		- - - - - - - - -		Post-Gondwana erosion begins
	MESOZOIC	Jurassic		Continental		Gondwana erosion begins F V
200		Triassic		Interclaire	Stormberg	F V
		Permian		- - - - - - -	Beaufort Ecca	
300		Carboniferous			Dwyka	Dwyka glaciation of Southern Africa
400	PALAEOZOIC	Devonian			- - - - - - - - - - -	
		Silurian			Cape	
		Ordovician			- - - - - - - - -	
500 600		Cambrian				
4,500	PRE-CAMBRIAN	Continental crust of Africa emerges				

1 Africa: Geology and Evolution

Africa is a huge continent covering 30·3 million km² or about twenty per cent of the earth's land surface. The continent extends some 8000 km through 72 degrees of latitude from Cape Agulhas in the south (34° 51′S) to Cape Blanc in the north (37° 51′N) and its greatest east-west length lies between Ras Hafun (51° 50′E) and Cape Verde (17° 32′W), a distance of 7200 km. Africa has over three-quarters of its area lying within the tropics and one third is affected by wind belts producing arid and semi-arid conditions.

Geological Structure of the Continent

The geological structure of this vast continent is relatively simple. Africa is basically a huge continental shield, which has existed since Pre-Cambrian times some 4500 million years ago. It is therefore somewhat similar to the smaller continental shield areas of the Baltic, Laurentia and Brazil. The continent has experienced several mountain building periods or periods of orogenesis and has been warped, folded and faulted. Some parts were folded more than 1500 million years ago and have since been little disturbed; these regions of stability are known as cratons. Regions which have been folded more recently—over the last 1200 million years—are called orogens or regions of orogenesis (Fig 1). The cratons contain most of Africa's gold, diamonds, chromite, asbestos and iron, and the orogens the continent's copper, lead, zinc and tin.

Pre-Cambrian rocks underlie the continent and are exposed over approximately one-third of Africa's surface (Fig 2a). Following the Pre-Cambrian Era there is little evidence of Africa's geological history during the early Palaeozoic Era (the Cambrian, Ordovician and Silurian periods) due to extensive erosion although some Cambrian shales, sandstones and limestones exist in the Sahara. In southern Africa the next important rock series after the Pre-Cambrian is of Devonian age and includes the folded shales and sandstones of the Cape range.

During the Mesozoic Era from the Carboniferous to the lower Jurassic period an important series of rocks was laid down called the Karoo System. These rocks were formed from materials eroded from the continental shield under climatic conditions ranging from desert to glacial. Karoo sediments once covered almost the whole of southern and central Africa from the Congo Basin to the Cape but they have been extensively eroded and now cover about half the surface of South Africa and other small areas (Fig 2b). Coal deposits at Wankie in Zimbabwe, in southern Zambia, and in the Transvaal indicate that humid tropical conditions prevailed at this time which gave rise to swamps and forests. At about the same time, the Nubian Sandstones of southern Egypt, the Sudan and Libya were forming and in West Africa clays and sandstones were being laid down during the Jurassic and Cretaceous periods. These West Africa rocks are known as the Continental Interclaire System.

During the Jurassic period, invasions by the sea along the margins of the continent caused extensive marine deposition especially along the eastern coasts (Fig 2b). The sea invaded again during the Cretaceous and Eocene periods in East, North and West Africa, producing vast thicknesses of sandstone and limestone; the northern deposits were later folded to form the Atlas ranges. At this time, much of southern Africa remained above sea-level, although some marine deposition occurred along downwarped parts of the coast as in Natal.

Gradually the seas began to retreat and the exposed rocks were subject to erosion during the Tertiary period. The wind-blown sands of the Sahara and Kalahari basin were formed at this time.

Our knowledge of Africa's geological history is therefore incomplete in many regions due to long periods of erosion. Africa is a continent which, since the Jurassic, has experienced only limited invasions by the sea and thus the thin layers of younger rocks covering the Pre-Cambrian base are largely continental and not marine in origin. As we shall see, warping, folding, faulting, vulcanicity and erosion have been the forces which have produced the present landforms and landscapes of Africa.

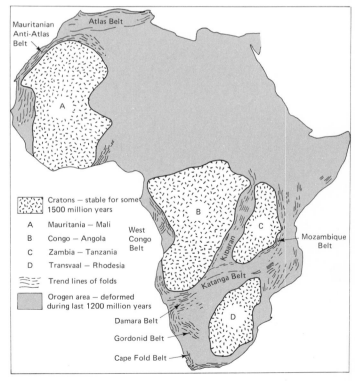

Fig 1.   The major structural regions of Africa

Fig 2 captions (left column)

a. Pre-Cambrian to end of Devonian

Beginning over 350 million years ago

Basement Complex: Archaean igneous and metamorphic rocks (granites, gneisses, schists) — the oldest rocks

Pre-Cambrian and Cambrian — less metamorphosed sediments (over 500 million years old)

Silurian and Devonian sedimentaries (400–440 million years old)

b. Carboniferous to end of Cretaceous

350 to 70 million years ago

Continental intercalaire (sandstones, conglomerates and clays)

Nubian Sandstones

Karoo beds — sedimentary rocks of continental origin formed in Carboniferous and Jurassic — 180 to 350 million years old. Lavas in black.

Marine sedimentary rocks (limestones, shales) laid down during sea invasions from Jurassic (180 million years ago) to recent times

Continental sedimentary rocks formed from wind-blown sands during Jurassic and Cretaceous periods 135 to 180 million years ago

c. Tertiary and Quaternary Periods

Beginning 70 million years ago

Volcanic rocks

Sedimentary rocks derived from weathering of uplifted parts of continent

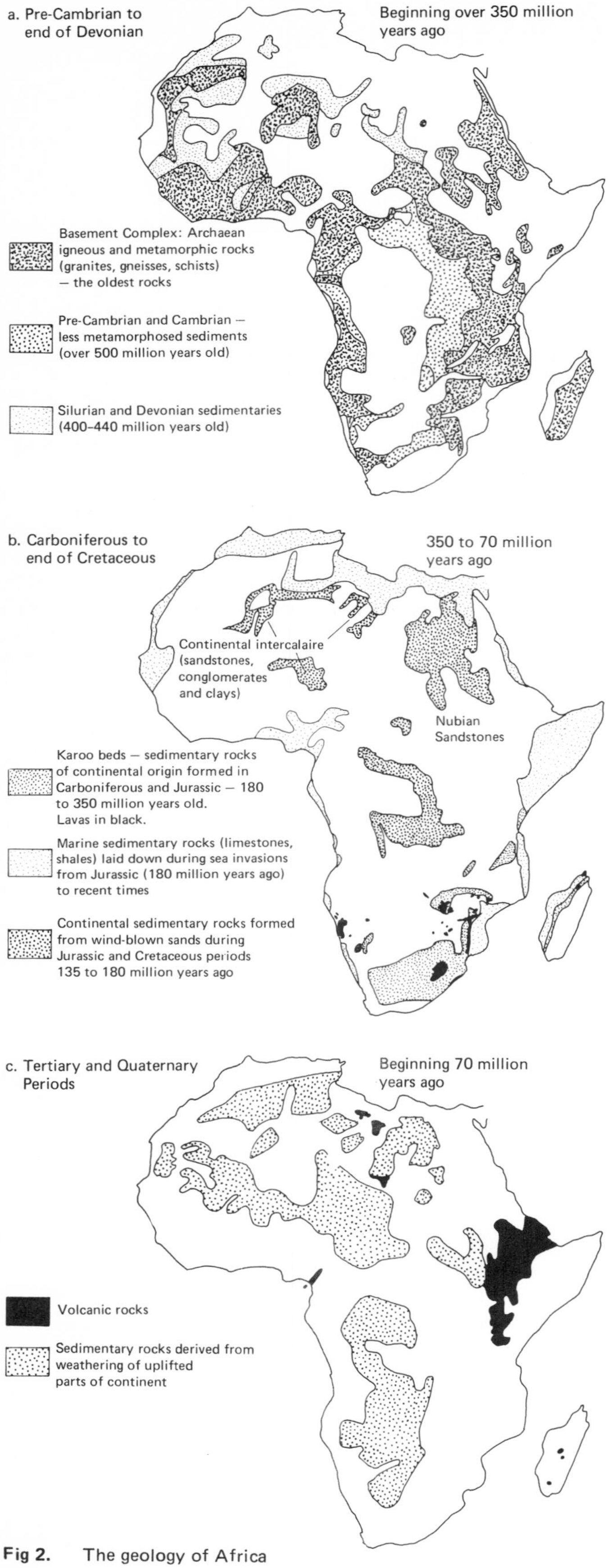

Fig 2.   The geology of Africa

The Emergence of Africa— Continental Drift

At the end of the nineteenth century, the Austrian, Edward Suess, produced the theory that the southern continents had once been part of a super-continent which he named *Gondwana-land* after the Gondwana region of India. This idea was expanded by the Englishman, F. B. Taylor, in 1910 and later by the German, Alfred Wegener, who stated that some 200 million years ago Gondwanaland was part of an even greater continent including North America and Eurasia which he named *Pangaea*. Pangaea, Wegener said, began to break up at the beginning of the Mesozoic Era (about 200 million years ago), the pieces drifting apart like solid rafts in a vast ocean of denser, more plastic rocks. To support his ideas, Wegener referred to the jig-saw fit of coasts, the similar rock sequences in the Karoo beds of southern Africa, the Deccan of India and the plateaux of Brazil, and the similarities of rocks and fossils throughout the southern continents. Most scientists of the time, however, could not accept this idea of lateral crustal movement since there was no known force which could move continents across the surface of the earth's dense basaltic crust.

Since the late 1950s, however, earth scientists or geophysicists have produced convincing evidence that continents really do drift. From this evidence it seems that Africa holds a key position since it remained relatively stationary as the other continents of Gondwanaland drifted away (Fig 3).

What causes the lateral movement of continental crust? In the early 1960s, the American, H. Hess, proposed the *theory of sea-floor spreading*. The earth's mantle, Hess stated, behaves somewhat like a giant convection system: material heated by

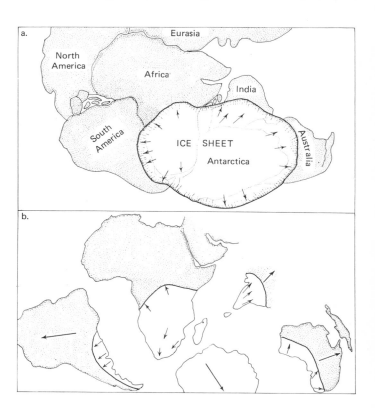

Fig 3.   (a) Pangaea and the extent of the ice sheet cover during Permian and Carboniferous times about 300 million years ago. (b) Regions of the present continents where evidence of the old glaciations exist. The arrows show directions of ice sheet movement

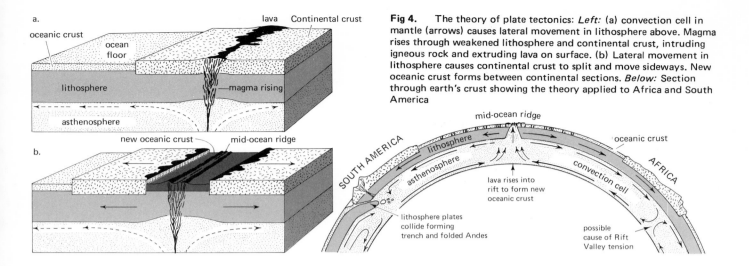

Fig 4. The theory of plate tectonics: *Left:* (a) convection cell in mantle (arrows) causes lateral movement in lithosphere above. Magma rises through weakened lithosphere and continental crust, intruding igneous rock and extruding lava on surface. (b) Lateral movement in lithosphere causes continental crust to split and move sideways. New oceanic crust forms between continental sections. *Below:* Section through earth's crust showing the theory applied to Africa and South America

radioactive elements in the earth's interior slowly rises in the asthenosphere—the softer plastic rock zone below the lithosphere (Fig 4). This magma reaches the surface along the mid-ocean ridges and flows away from them, cooling and hardening to form the rigid lithosphere. New lava emerging from the ridges attaches itself to the rear of the solidified older lava plates and forces them along laterally. After millions of years the lithosphere plates will have moved thousands of miles by constant addition of new lava at their rear. The leading edges of the plates will eventually be forced down into the asthenosphere by sinking under a continental crustal block thus forming deep ocean trenches along the edge of the continent. Crumpling of the continental surface will occur along these zones creating folded ranges (Fig 153, p. 82). Where the surface plates are moving apart the action will be reflected at the surface by zones of fracturing, tension, faulting and earthquakes as, for example, in the Rift Valley region of East Africa. This is basically the *theory of plate tectonics.*

The theories of sea-floor spreading and plate tectonics have been supported by convincing evidence. Seismic shocks, for example, are more frequent at the troughs and the mid-ocean ridges than anywhere else, that is, where the plates are sinking or rising. Again, rock core samples drilled from the ocean beds show that rocks increase in age away from the oceanic ridges, just as one would expect. Moreover, it is known that iron particles in rocks align themselves with the earth's magnetic axis, now roughly north–south. In the last 76 million years the earth's magnetic field has reversed itself 1717 times and, if the floor-spread theory is correct, then each new magnetic change would be recorded in the rock particles outwards from the oceanic ridges. Magnetic readings taken from the ocean beds confirm a zig-zag orientation of rock particles.

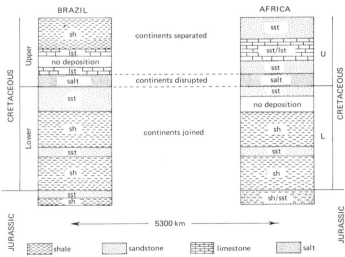

Fig 6. The sequence of sedimentary rocks of Jurassic and Cretaceous age as shown by boreholes in the basins of northeast Brazil and the Cameroon region of Africa

If continents do move, then there must be evidence that they were once joined together. The case of Africa's link with South America is supported by the following:

1. Visual fit: The east coast of South America and the west coast of Africa have good visual fits, not only at the surface, but also at 1000 m and 2000 m depths.

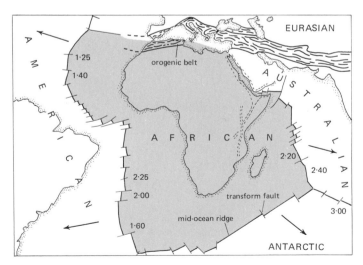

Fig 5. The African Plate is one of six major plates which form the earth's surface. The plates are moving continuously in the directions shown. The figures show the rate of movement in centimetres per year

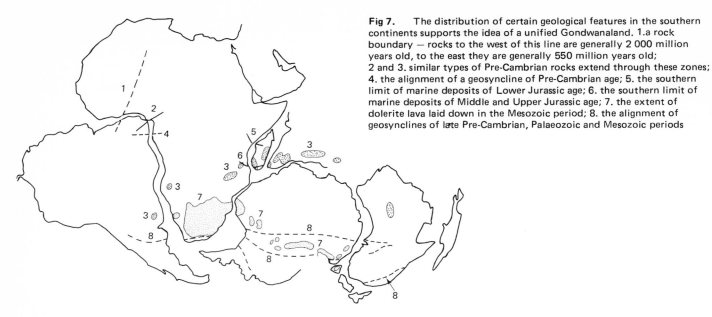

Fig 7. The distribution of certain geological features in the southern continents supports the idea of a unified Gondwanaland. 1.a rock boundary — rocks to the west of this line are generally 2 000 million years old, to the east they are generally 550 million years old; 2 and 3. similar types of Pre-Cambrian rocks extend through these zones; 4. the alignment of a geosyncline of Pre-Cambrian age; 5. the southern limit of marine deposits of Lower Jurassic age; 6. the southern limit of marine deposits of Middle and Upper Jurassic age; 7. the extent of dolerite lava laid down in the Mesozoic period; 8. the alignment of geosynclines of late Pre-Cambrian, Palaeozoic and Mesozoic periods

2. Geometric fit: If two identical shapes are drawn on the surface of a sphere, they can be brought into coincidence with each other by rotating them towards each other about a fixed axis. The west coastline of Africa and the eastern coastline of South America fit almost exactly on each other if rotated through an angle of 57° with the rotation point at 40°N and 30°W. Overlaps, as in the Niger delta, may be disregarded since such features formed after the Gondwanaland break-up.

3. Matching geology: Both South America and Africa are composed of rocks of varying ages and there is a convincing boundary join across the two continents between Accra and São Luis in Brazil dividing pan-African rocks (550 million years old) and Eburean rocks (up to 2000 million years old).

4. Matching of orogenic zones: The alignment of belts of fold mountains (orogenic zones) matches across the join of the two continents. Folded ranges in the Falkland Islands and Argentina are of similar age and structure to those of the south-west Cape in South Africa.

5. Glacial evidence: In eastern Brazil, Paraguay and Argentina lie thick deposits of tillite, a fossilised glacial moraine (p. 74), exactly like those of southern Africa. In some places the deposits are over 1000 m thick. It is most likely that they are the deposits of a vast Gondwanaland ice sheet which covered the united continents of Africa and South America (Fig 8).

6. Sedimentary basins: Along part of the north-eastern Brazil coast, south-eastern Nigeria and Cameroon, similar sedimentary rock sequences exist, the beds created in shallow basins of Cretaceous and Jurassic age. The lower beds match exactly on both continents with salt deposits of upper Cretaceous age overlying them suggesting an invasion of the sea as the continents split (Fig 6). The sequence above the salt beds, however, ceases to match as the rocks developed in differing environments. From this it is estimated that the continents broke up during the lower Cretaceous about 100 million years ago.

7. Palaeomagnetic evidence: The fact that rock particles have magnetic properties allows geophysicists to reconstruct the position of the poles in past geological times and also the probable climatic belts of the past. From this, it appears that southern Africa and South America lay within the Arctic Circle of Permian and Carboniferous times and that during the Triassic the continents had moved some 40° closer to the equator.

Similar techniques have been applied to the eastern coasts of Africa and good visual fits at 100 m depth obtained. Large gaps

in the Red Sea and south of Madagascar may be due to the sinking of large parts of Gondwanaland. In 1974 the *Glomar*

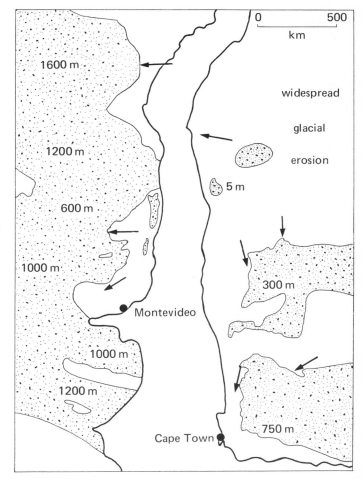

Fig 8. The distribution and thickness of glacial deposits of Permian and Carboniferous times in Africa and South America. Figures indicate thicknesses, arrows indicate direction of ice movement

Challenger, a U.S. research vessel, discovered large sections of continental crustal granite aged about 600 million years on the ocean floors. This is probably an extension of the undersea Falkland Plateau and would fit the gap which had previously existed between southern Africa and Antarctica.

Experiments

Fitting the continents together:

1. Obtain a large world globe which shows submarine contours. Trace the western coasts of Africa and the eastern coasts of South America and along submarine contours shown. Bring shapes together to see which gives the best visual fit.

2. Place a piece of tracing paper on the globe and trace on to it the eastern coast of South America. Locate the rotation pole (the Azores Islands), and rotate the tracing through an angle of 57 degrees towards Africa. Write up your findings.

Questions

You should take about 35 to 40 minutes to answer questions in this book, building your answers round your maps and diagrams.

1. Write a short account of the geology of Africa and relate the distribution of Africa's minerals to cratons and orogens.

2. Outline the theory of continental drift as proposed at the beginning of this century. What arguments would you use to support this theory from your personal observations and from the experiments of others?

3. With the aid of a large, clear sketch-map, analyse the geology and structure of your own country.

2 Africa: Surface Morphology and Drainage

Morphology

Africa's surface is generally one of level, extensive plateaux, highest in the east and south, lowest in the north and west. Nearly half of the continent lies above 1000 m and nearly two-thirds above 350 m. Four-fifths of the regions lying to the south of the equator lie above 350 m, partly due to the tilting of the continent towards the northwest. The continent's monotonous level surfaces vary in altitude from 600 m to 2600 m and are the result of long periods of erosion separated by periods of slow uplift of the continent.

This continental uplift was highest in the southeast where the rim of Africa rises to between 1500 m and 2100 m in the Maluti and Drakensberg mountains (Mont aux Sources, 3280 m). This high rim is continued northwards into the Eastern Highlands of Zimbabwe in the Chimanimani Mountains and the Inyanga escarpment (Mt Inyangani, 2953 m). In East Africa the crust was upwarped into a huge dome to between 3000 m and 4000 m in the region of the Kenya Rift Valley (Settima, 3999 m) and to a general plateau level above 1500 m. In the northeast of the continent the great massif of the Ethiopian Highlands rises abruptly to rolling uplands standing between 2000 m and 2400 m, overlooked by mountain peaks exceeding 4500 m.

In the western regions of Africa south of the equator, the surface is generally lower but still displays some impressive relief in the Central Highlands of Namibia which rise to over 2400 m in the Brandberg (Königstein Peak, 2586 m) and the Bihé Plateau of Angola (highest point 2619 m).

A line drawn along the western edges of the Ethiopian Plateau, the rifted rim of the Congo basin, through Kamina to Luanda, would divide this higher part of Africa from the lower regions to the north and west which lie generally below 1000 m. In West Africa few regions are above 600 m except the elevated Adamawa Highlands in the Bamenda region (2679 m), the Jos Plateau (1780 m), the N'Zérékoré of northern Liberia (1752 m), and the Futa Djallon Plateau (1425 m). Northwards, the sands of the Sahara bury the rock shield of Africa.

These landscapes of monotonous erosion levels separated by steep, low escarpments are interrupted by various landforms due either to planation itself or to vulcanism, warping, folding, or faulting. The isolated mountain or inselberg is a remnant feature of deep weathering and erosion (p. 29) and its solid bulk often juts up abruptly from the African plain, dominating the landscape. Inselbergs are common in the drier tropical regions of central and northern Nigeria, Tanzania, Kenya, Zimbabwe and Malawi but are also seen in the humid tropics and in the subtropical regions of South Africa.

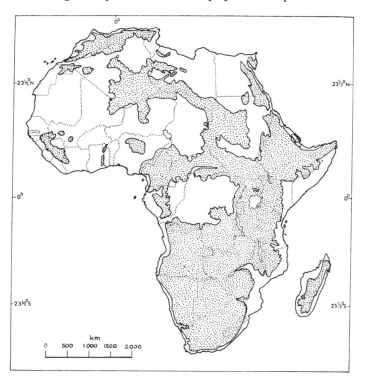

Fig 9.   Africa: Land over 450m above sea-level

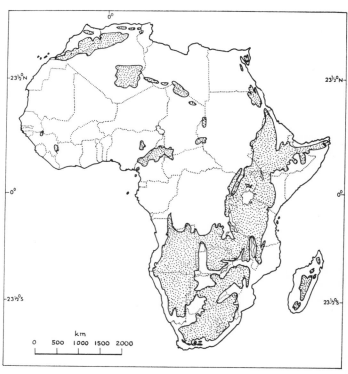

Fig 10.   Africa: Land over 900m above sea-level

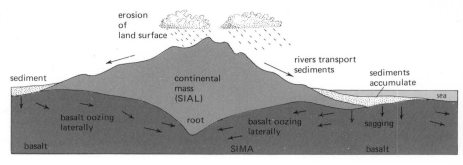

Fig 11. The theory of isostatic readjustment. The continent will slowly rise to compensate for the loss of material by erosion

The Effects of Vulcanism, Folding and Isostatic Uplift

Volcanic activity has had a major influence on the development of Africa's landscapes. In Ethiopia, lava flows 500 m thick have buried the crystalline plateau and in East Africa four volcanic peaks rise to over 4000 m, including Kilimanjaro whose ice-capped summit, Kibo (5895 m), is the highest in Africa. Further south, lava cappings have increased the height of the Drakensberg by some 1500 m and the Lebombo Mountains by 1000 m. In West Africa, active Mt Cameroon (4070 m), even though it lies only $4\frac{1}{2}°$ north of the equator, is often snow-capped. The Bamenda Highlands were also buried under thick masses of lava. Further north, the low plateau of the Sahara is broken by volcanic massifs and the crystalline rocks of the Hoggar,

once forming a plateau 2000 m high, are now overlain by lavas 200 m thick and overlooked by high peaks rising to 2198 m in Tahat. Dark basaltic lava flows and volcanic pinnacles form the rugged landscapes of the massifs of Aïr (1900 m) and Tibesti (Emi Koussi, 3415 m).

In the north and south of the continent folding of sedimentary strata against the rigid block of Africa has produced a very different landscape from that of the interior plateaux. Ghana and Togo also have remnants of an ancient folded range in the Akwapim-Togo-Atacora mountains. In the northwest, the Atlas ranges stretch for some 500 km from east to west and rise to over 4000 m in the High Atlas. These mountains were formed from huge thicknesses of sedimentary rocks, chiefly sandstones and limestones laid down in a vast depression between the continental blocks of Europe and Africa. Pressure from the European block caused intense folding in the northern sections of this orogenic belt, but the pressure lessened as the sedimentaries splayed out on to the African shield. The Cape Ranges of South Africa are lower and more ancient, their origins being traceable to the late Triassic period (p. 83).

The isostatic uplift of Africa did not occur evenly and some areas tended to sag and form saucer-like shallow basins separated by upwarped parts of the plateaux, fault blocks or mountains (Figs 13 and 14). Powerful forces originating deep down in the earth's crust may have helped this downwarping process. The shallow basins became the reception areas for networks of rivers whose waters created vast lakes like that of Lake Congo (p. 68). The increasing weight of lake and river sediments caused the basins to sink further and this produced minor fracturing and faulting along the basin edges. The largest of these basins, the Congo, covering an area roughly equal to half Australia (4 million km^2), is extremely old, originating about 400 million years ago and its vast level floor was covered by thick deposits of Karoo sediments. Other younger basins include Chad (p. 68) which covers 2·5 million km^2 and was formed during the Cretaceous period, also El Djouf, Sudan, Kalahari and the more recently formed Lake Victoria-Kyoga basin associated with Pleistocene warping and rifting.

The giant lakes which occupied these shallow depressions increased in size during long rainy periods, the pluvial periods. In the case of the Congo and the Victoria lakes, overflow rivers cut through the rims of the basins, draining Lake Congo entirely and leading Lake Victoria's waters into the Nile system. Other basins appear to have extended beyond Africa's present coasts and probably into adjoining land masses before the break-up of Gondwanaland. The Orange basin may be an example of this for the river flows westwards in contrast to the easterly flowing Limpopo and Zambesi rivers. With the break-up of Gondwanaland such lakes would be partially or completely drained.

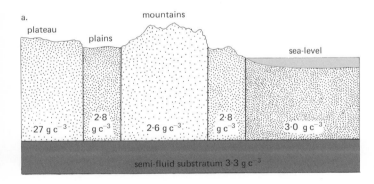

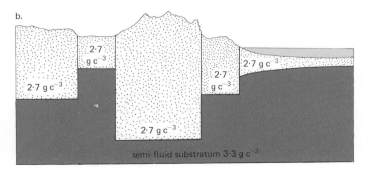

Fig 12. The theory of isostacy: two ideas. (a) The lithosphere consists of blocks of differing densities; each rises to a height depending on its density. (b) The lithosphere has the same density throughout but varies in thickness. High mountain ranges have roots in the substratum, plateaux are thinner

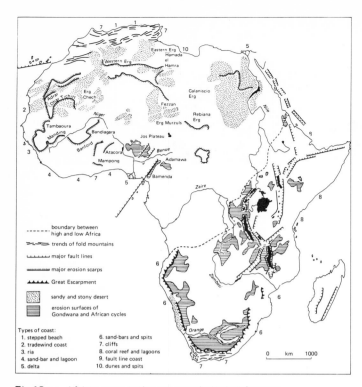

Fig 13.   Africa: some major geomorphological features

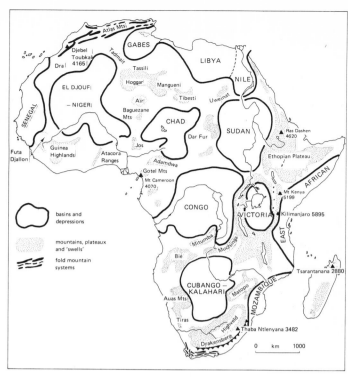

Fig 14.   Africa: major structural basins, depressions and divides

The Major Drainage Systems of Africa

The major drainage patterns of the African continent have been strongly influenced by the 'basin and swell' nature of Africa's surface and the nature of the major rivers reflects this influence.

The Nile is a very old river dating from the Eocene period about 65 million years ago and has experienced many changes of course, although it has always flowed to the north. During the pluvial periods of the Pleistocene the area at present occupied by the swampy sudd between Juba and Khartoum was once a vast lake, called Lake Sudd. The lake was fed by ancestor rivers of the present White and Blue Niles and eventually the waters overtopped the rim of the basin and flowed northwards to the Mediterranean through the Sabaloka gorge 150 km north of Khartoum. Lakes may also have formed along the ancient Nile course between Malakal and Khartoum as shown by the shore deposits well above the present river and by widespread deposits of lake sediments. These ancient lakes may have rivalled in length and width the present-day lakes Tanganyika and Malawi.

The Nile, 6530 km long, and its major tributaries, the Sobat, the Blue Nile and the Atbara, dominate the drainage of the whole of northeastern Africa. The Sobat rises in the southwest of Ethiopia and drops rapidly by a series of falls to enter the eastern part of the great Sudd region. Here its flow is sluggish due to the thickness of the vegetation, its maximum flow being felt in the lower regions in November. Its flood waters pond back the flow of the White Nile in the Bahr el Ghazel swamps and this results in much water loss by evaporation. The Blue Nile rises in the Gojjam Highlands, flows northwards into the lava-dammed ice-blue Lake Tana which acts as a regulating reservoir, then cuts down through steep gorges by a series of

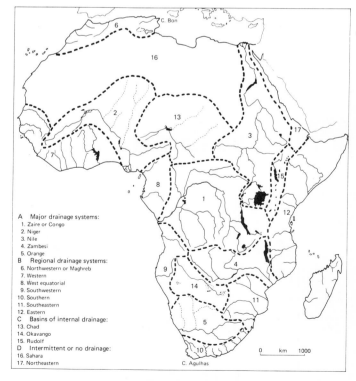

A   Major drainage systems:
  1. Zaire or Congo
  2. Niger
  3. Nile
  4. Zambesi
  5. Orange
B   Regional drainage systems:
  6. Northwestern or Maghreb
  7. Western
  8. West equatorial
  9. Southwestern
  10. Southern
  11. Southeastern
  12. Eastern
C   Basins of internal drainage:
  13. Chad
  14. Okavango
  15. Rudolf
D   Intermittent or no drainage:
  16. Sahara
  17. Northeastern

Fig 15.   Africa: the regional systems of drainage

falls and rapids. Its highest level is in September although it rises throughout July and August. The Blue Nile's contribution is the main cause of the summer floods of the Nile in Egypt. The Atbara joins the Nile some 320 km below Khartoum and increases the water supply during late summer and early autumn but during the rest of the year it is merely a line of disconnected pools. The tremendous erosive power of these rivers is seen in the huge amounts of silt they have deposited along the banks of the lower Nile and in the huge delta (p. 53).

The Niger is an unusual river since its upper course has all the features of a complete river from a vigorous active flow to a winding and sluggish flow, while the lower course, except for the section which crosses the Sokoto Plains, flows in a clearly cut valley. These changes are explained by the river's complicated history. In the Pleistocene period the present lower Niger was an independent river flowing southeastwards, fed by streams rising in the southern Hoggar and by the Sokoto and Benue (Fig 16a). The upper Niger, rising on the Futa Djallon plateau, flowed westwards to the Atlantic during the late Pliocene and early Pleistocene pluvial periods. In a drier interpluvial period 10 000 to 15 000 years ago this outlet was blocked by drifting sand dunes of the Ougadougou sand sea and the Niger was diverted eastwards into a large basin to form L. Arouâne. The waters of this ancient lake rose until they reached the headwaters of the lower Niger which drained the lake and created the present course (Fig 16b).

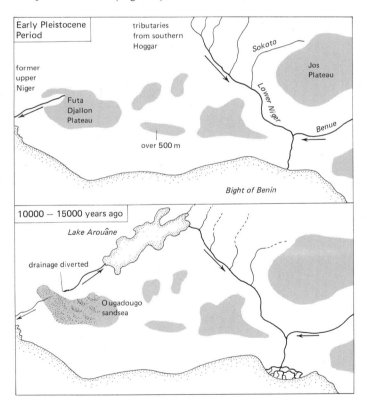

Fig 16.   Stages in the evolution of the Niger drainage system

The Niger now rises at 800 m in the Futa Djallon only 250 km from the Atlantic coast but flows eastwards in a great arc 4160 km long. Its upper course is joined by rushing tributaries on the plateau and between Bamako and Koulikoro it crosses 60 km of rapids before gently dropping over 300 km to the flood plains of the Inland Delta. These plains are formed from the sediments of former L. Arouâne and cover 17 000 km^2. Here the

river is joined by the Bani which rises in the highlands of the Ivory Coast. The Niger is highest here during August, following the July rains in the highlands. The river leaves the Inland Delta via the Tosaye sill through which L. Arouane overflowed to join the lower Niger in the mid-Quaternary. The river then cuts through the crystalline basement and flows over several rapids at Fafa, Labbezenga and Bussa. From Jebba onwards the high water period extends from August to mid-December. The Niger then breaks up into a number of distributaries flowing across its huge coastal delta.

The Zaïre or *Congo River* occupies the basin once partly covered by L. Congo. There may have been an older Congo river whose outlet to the Atlantic was blocked by the steady uplift of the rim of Africa. Before the Pleistocene, a coastal river had succeeded in eroding backwards through the rim and drained off the lake waters. The remnants of ancient L. Congo are seen in the shallow lakes Upemba, Maindombe and Tumbo and their marshes in the southwestern part of the basin. Although the climate became much drier some 10 000 to 15 000 years ago, the Congo/Zaïre River maintained its course throughout the Pleistocene, establishing the pattern we see today.

The Zaïre with its upper section, the Lualaba, is now 4350 km long and curves in a great arc through the northern parts of its basin before cutting through its gorge in the Crystal Mountains. Over a distance of 2000 km between Kisangani and Kinshasa, the river falls only 100 m; here its width varies from 3 km to 15 km and it is slow-flowing with numerous braided channels. Just below Stanley Pool at Kisangani where the original point of capture by the coastal river lies, the Zaïre drops 275 m over 32 separate rapids in the 350 km to Matadi.

The River Zambesi, 3000 km long, rises at 1365 m in Zambia's extreme northwestern tip, flows across the flat Kasisi Plains, then crosses Karoo basaltic lavas in a series of rapids before plunging over the Victoria Falls below which it enters its zig-zag gorges (p. 59). It then enters Lake Kariba and below the Kariba Dam is joined by the Kafue and the Luangwa. It again flows through a narrow gorge at Quebrabassa, the site of the Cabora Bassa dam whose lake now drowns the river's lower middle course. Here the Zambesi is only 800 m wide. The course ends in the fairly small estuarine delta on the Mozambique coast (Fig 88).

It is thought that the upper sections of the Zambesi did not always drain eastwards but, with the Luangwa and the Orange, flowed westwards into the Kalahari Basin (Fig 17). The floor of the Kalahari Basin was raised by the vast quantities of alluvium deposited by these rivers and a huge lake formed whose waters overflowed eastwards. During the Cretaceous period, Madagascar drifted away from southern Africa and the lower middle course of the Zambesi was down-faulted in a deep trough. Water from L. Kalahari flowed along this trough drawing the upper Zambesi, the Luangwa and the Kafue eastwards to the Indian Ocean. The Orange-Vaal system, flowing westwards, ceased to receive water from the ever-dwindling L. Kalahari.

The River Orange today rises only 160 km from the Indian Ocean but flows 1100 km to the Atlantic, its course length being 2000 km. The Orange drains some 855 000 km^2 and rises in the Maluti Mountains of Lesotho. From this high rim of Africa, the Orange drops 3350 m to the interior plateau at 1300 m where, 450 km from its source, it is joined by its largest tributary, the Vaal. The Orange then zig-zags through a region where annual rainfall drops from 650 mm to 250 mm a year. From Prieska the river falls only 600 m in 400 km in a region where rainfall averages 125 mm a year and annual evaporation rates exceed 2500 mm. The river then drops over the Aughrabies

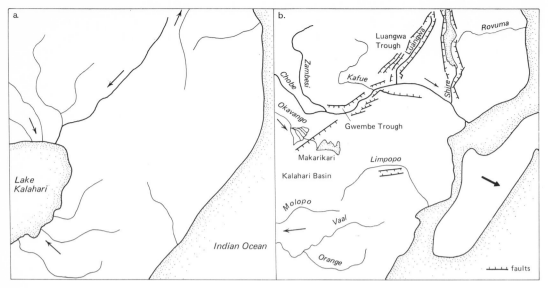

Fig 17. Stages in the evolution of the Zambesi drainage system

Falls (Fig 102) and flows through arid landscapes for nearly 500 km before entering the Atlantic.

The Coastal Fringes

Since much of Africa has been uplifted there are few extensive coastal lowlands except for the plains of Mozambique, those of northeastern Kenya and southern Somalia, and the coastal zone of Senegal and Mauritania. The continental shelf of Africa is widest in the Gulf of Gabes where depths of less than 200 m extend for up to 400 km from the shores and for up to 240 km out from the South African coast on the Aghulas Banks. But in general the continental shelf is narrow, usually less than 50 km, and off Natal and Angola it is only 5 km wide. This narrow shelf separates the basaltic ocean floors from the granite platform of Africa.

The coast itself lacks major indentations and its length of 27 000 km is only half that of Asia. The coasts are generally low and sandy and there are few good natural harbours except where downwarping or local faulting and sinking has produced deep inlets, e.g., at Freetown and Mombasa. Deep river estuaries are few, the only major examples being at the mouths of the Gambia and Congo rivers.

Questions

1. With the aid of a large clear sketch-map, outline the surface morphology of Africa.

2. How far is it true to say that the landscapes of Africa are dominated by monotonous erosion levels?

3. In what ways has the basin and swell surface of Africa affected the drainage pattern of the continent? Confine your discussion to the major rivers only.

4. Explain, with the aid of sketch-maps, the unusual courses taken by the Nile, the Congo, Zaïre and the Zambesi.

3 Rocks and Weathering

The Nature of Rocks

Rocks are composed of minerals of which the most common are the iron oxides (4% of rock minerals), calcite and dolomite (9%), kaolinites or clay minerals (18%), quartz (28%) and the feldspar group (33%). Rocks can be distinguished by their origin: igneous rocks are formed from cooled, solidified magma or lava; sedimentary rocks from weathered particles; metamorphic rocks have altered structures and compositions. About 75 per cent of the total surface of the continents consist of sedimentary rocks while the remaining 25 per cent is of crystalline (igneous and metamorphic) origin. The sedimentary rocks, however, form only a thin covering of the earth's crust and represent about 5 per cent of the lithosphere, the remainder being crystalline. The sedimentary shales, sandstones and limestones and the igneous granites and basalts are the most common landscape-forming rocks.

1. Igneous rocks are derived from molten materials from beneath the earth's crust. They are crystalline, the size of the crystals depending on the rate of cooling—the longer the cooling, the larger the crystals. Plutonic igneous rocks have cooled extremely slowly at great depths and have large crystals over 1·25 mm long. Rocks formed from swiftly cooling surface lava contain much smaller crystals (0·5 mm). Igneous rocks which have cooled near to the surface are called hypabyssal rocks and have either medium-sized crystals or large crystals set in beds of finer crystals.

Igneous rocks may also be classified as acidic, intermediate or basic according to the proportion of rock-forming minerals— quartz, feldspar (plagioclase and orthoclase), biotite (a dark mica), augite, hornblende, olivine or dioxide. Acidic igneous rocks have high amounts of quartz and feldspar, basic rocks contain much mica, augite, hornblende and olivine, while intermediate types have a mixed mineral structure. Although several hundred types of igneous rock are recognised, only some nine or so are really common; these are classified in the Appendix.

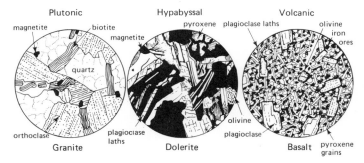

Fig 18.   Mineral composition of three igneous rocks as seen under a microscope. Magnification about x 25

2. Sedimentary rocks are composed of deposited minerals and rock fragments produced by mechanical and chemical weathering of former rock masses or by organic action. The materials are squeezed, cemented and hardened by pressure of overlying beds, the cements being formed from minerals surrounding the grains of rock or the partially dissolved grains themselves. Silica, carbonates and iron oxides are the chief cementing agents.

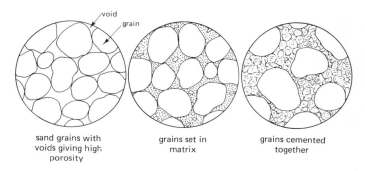

Fig 19.   Three types of sandstone as seen under a microscope. Magnification about x 25

The general texture of sedimentary rocks depends on the circumstances under which they were laid down. Thus wind-sorted grains produce rocks of a similar grain size and texture while ice-sorted material forms rock of irregular grain size and coarseness. A close examination will often reveal a rock's origin. Sandstones, for example, contain quartz grains which indicate a river origin if sharp and angular or a wind-blown or aeolian origin if rounded and smoothed. Again, shales result from the consolidation of clay and under such pressure shale forms distinct layers which cause the rock to split easily. Limestones and dolerites may contain tiny shell fragments while a fourth group, evaporites, are caused by intense seawater concentration and the formation of minerals such as gypsum, salt and potassium salts. Such minerals may flow under pressure and intrude into other rocks, e.g. in the form of a salt dome.

Over relatively short distances the texture of sedimentary rocks of the same age may change from coarse to fine. Lighter particles will have been carried further by water currents before they settle on a lake or sea bed. For example, shales may have been derived from finer material such as silt some distance from an ancient shoreline, sandstones just offshore, and conglomerates (coarse pebbles and stones set in finer material) near the beach. Table 2 in the Appendix classifies sedimentary rocks according to their mode of formation and their chemical origin.

3. Metamorphic rocks are produced by alteration of pre-existing rock. Shales change to schists, limestones to marbles, sandstones to quartzite, and igneous rock to granulite. The change may be caused by intense heat created by an igneous intrusion (thermal or contact metamorphism), or by heat and pressure created during rock movements, e.g. during folding or faulting, by stress due to pressures and shearing during mountain building (regional metamorphism), or by the movement of fluid chemical elements in the rock. Metamorphism recrystallises minerals into larger grains, alters and rearranges the grains, and further combines chemicals to create new minerals.

In the aureole (the zone affected by heat around an igneous intrusion) thermal or contact metamorphism causes minerals such as clays to disappear while other minerals such as biotite mica expand and develop larger crystals. Rocks harden, crystals become coarser, and the rocks are able to resist further high temperatures.

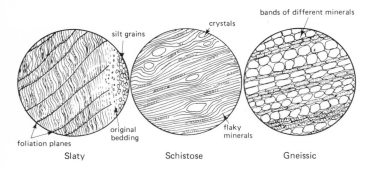

Fig 20. Three types of metamorphic rocks as seen under a microscope. Magnification about x 25

In zones where mountain building has occurred, regional metamorphism subjects rocks to great pressures over wide areas. The rocks often yield by folding and faulting. New minerals and mineral rearrangements occur according to the degree of pressure and heating. A mudstone, for example, will change to a slate under moderate pressure; with increasing pressure the slate will change to a schist with a silvery green surface; at high pressure schists such as garnet or mica may develop. Foliation occurs where a mineral such as quartz forms banded layers a millimetre to several centimetres thick. Such slaty schistose and foliated rocks allow penetration of water and are the more easily weathered. (See Appendix for a classification of metamorphic rocks.)

Weathering

Weathering is the process by which rocks are gradually broken down into increasingly smaller particles and thus prepared for removal by wind, ice and running water. The process is the first stage in the wearing down of the land surface—the cycle of denudation—and provides soluble and insoluble materials for transport and deposition.

Weathering may be mechanical, chemical or biological. Mechanical weathering reduces rock to increasingly smaller pieces, sometimes to tiny mineral grains, without changing the chemical composition. Chemical weathering causes chemical reactions and decomposition of rock minerals and produces new compounds. Biological weathering is the effect which organisms (plants and animals) have on rock disintegration.

The rate and type of weathering depends on several controls. The amount and duration of precipitation (rain, snow, hail, frost) controls the supply of a vital agent in chemical processes—water—while high temperatures increase weathering rates. An important organic control is the extent and type of vegetation cover: savanna, for example, allows a higher surface run-off of rainwater than tropical rain forest which hinders run-off and promotes deep vertical rotting of rocks by water seepage. Gentle or flat ground surfaces also encourage vertical seepage and deep rock weathering while different rocks will weather at different rates under the same climatic conditions. Weathering processes may also be interrupted by climatic changes from pluvial to arid conditions or by tectonic changes (faulting and folding), and by earthquakes or volcanic activity which may disturb weathering penetration.

Mechanical Weathering

Mechanical weathering includes the action of thermal expansion, pressure release, and crystal growth upon rocks causing them to break up into smaller pieces until a rock waste or debris of the same composition as the original rock is produced. Thermal expansion or insolation weathering causes fracturing of rocks by stresses set up by alternate heating and cooling. This is only really effective when accompanied by water: rocks heated to $210°C$ in laboratory experiments and then allowed to cool have shown no significant effects and only a slow change occurred at temperatures exceeding $400°C$. When the rocks were cooled by water, however, rapid disintegration occurred. Thus, a sudden rain shower on a hot sunbaked rock surface would set up stresses and create tiny surface cracks or microfissures which prepare the surface for deeper penetration by rainwater. No doubt thousands of years of temperature variations on their own may cause some kind of lowering of resistance (called rock fatigue) to weathering.

Heating and sudden cooling by rain showers causes exfoliation. When the rock surface becomes intensely heated it expands slightly against the cooler interior. A thunder shower may cool the thin surface layer rapidly to cause rapid shrinkage of rock minerals. Constant repetition produces microfissures and cracks between the surface layer and the cooler rock mass beneath. Curved plates or rock a few centimetres thick break away leaving new rock surfaces exposed to the same process.

Pressure release or unloading occurs where the rock mass has been exposed by the removal of overlying rock beds or debris. The gradual release of pressure causes the rocks, particularly granites, to expand, and curved joints parallel to the rock surface are formed. Rainwater penetrates along these joints, weathering proceeds, and curved masses of rock 'shell' slide or fall away. These fragments, varying in thickness from 0·5 m to 1·5 m, may be further weathered by exfoliation and chemical processes at the foot of the rock slope.

Granular disintegration caused by water being converted into ice (freeze-thaw weathering) is particulary active in well

Fig 21. Nigeria, the Jos Plateau: pressure-release sheeting on a domed inselberg near Panyam

jointed or porous rocks in regions where temperatures fall below freezing point, usually at night at high altitudes and during cold weather. Water contained in cracks and joints increases in volume by 9 per cent when converted to ice and exerts pressures approaching 150 kg cm^{-2} to cause rock disintegration, a process also known as frost riving. Rock pieces and individual grains are prised loose and prepared for wind or water erosion.

Rock distintegration may also be caused by the enlargement of salt crystals between rock grains when the crystals are moistened. This frequently occurs in semi-arid regions where salts are not readily washed away. Heating and cooling and the occasional damping also cause the physical expansion of salt particles creating stress and loosening surface grains.

Chemical Weathering

In chemical weathering the gases of the atmosphere—water vapour, oxygen and carbon dioxide—penetrate the rock and react in the presence of water with the rock chemicals to create entirely new chemical compounds. Thus the whole structure of the rock is attacked causing rock decomposition or rock rotting.

Pure water alone will easily dissolve salts in rock and these are swiftly removed in solution by surface wash. But rain water, as it passes through the atmosphere, absorbs some carbon dioxide and becomes a weak carbonic acid, a most effective agent of chemical weathering. Some humic acid, derived from rotting plants, bacteria and animal life, may be absorbed by rain water as it passes through the soil. Reaction between the various chemical compounds in the rock and slightly acidic rain water—hydrolysis—leads eventually to the complete dissolution of the minerals and a breakdown of rock structure. (The chemical formula for hydrolysis is $H_2O = H_+ + OH^1$.) In a rock such as granite, soluble salts, e.g. sodium and potassium carbonate, will dissolve first and be washed away, followed by the swift weathering of biotite. Then the feldspars break down into clays while quartz crystals resist to form loose sands.

Silicates, the most important single group of rock minerals, when combined with oxygen atoms in solution may be replaced by other mineral elements such as aluminium, iron, magnesium, calcium, sodium or potassium. In the tropics, silicate minerals are converted to clay, iron and aluminium compounds until the rotted rock layer or regolith is dominated by quartz or fine clays. Calcium, sodium, magnesium and potassium soon disappear under chemical weathering while iron compounds and aluminium remain as residual products. Iron compounds in the rock react with oxygen in water to form iron oxides and an oxidised rust-like crust forms over the rock surface which easily crumbles.

Some rocks are very prone to weathering and are termed unstable. Limestone is an example of an unstable rock. When weak carbonic acid in rainwater contacts limestone it converts insoluble calcium carbonate into soluble calcium bicarbonate which is soon removed, a process known as carbonation.

The rate of chemical weathering processes increases with a rise in temperature; chemical reaction rates may be doubled or even trebled with each 10°C rise. In the hot, wet tropics where ground temperatures are 10°–15°C higher than those of temperate regions, chemical weathering continues throughout the year and the rate of rock rotting is fast—three to four times that of temperate latitudes and ten times that of the tundra.

Biological Weathering

Organisms, both plant and animals, contribute to the surface weathering of rocks. Plants aid mechanical and chemical weathering by causing rocks to disintegrate under the pressure

Fig 22.   Uganda-Zaire, the Ruwenzori horst: jagged peaks produced by freeze-thaw shattering

of their growing root system and by adding chemicals such as nitric acid, ammonia and carbon dioxide to the soil. Leaf litter adds minerals to the upper regolith; in tropical rain forests, for example, up to 20 tonnes of dry matter containing almost 0.5 tonne of silica may be added to the regolith each year. Lichens, mosses and algae clinging to rock surfaces help to retain moisture and activate organic acids. Soil algae are microscopic plants containing chlorophyll and there may be up to 100 000 per gramme of soil. Humic acids from decayed plants help in the solution processes of iron and other minerals.

Soil bacteria and other minute organisms also assist in regolith decomposition. Bacteria are one-celled organisms and there may be up to 4 billion of them to 1 g of soil. Other animal life on a microscopic scale include protozoa, slightly

Fig 23.   Zimbabwe, near Bulawayo: disintegration of granite by chemical weathering along joints and on rock faces

larger than bacteria, and nematodes, larger and more complicated but still too small to be seen without a microscope.

Earth worms, ants, termites, spiders and snails are also active. Earth worms, of which there may be up to 2·5 million per hectare, can raise up to 45 tonnes of soil per hectare in the form of fine surface casts and their burrows allow freer passage of water and air through the soil. They also mix soil from lower levels with that on the surface. Termites create huge lime-enriched termitaries from soil brought from lower depths and from surrounding areas. Burrowing animals such as rodents, rabbits, ratels or honey badgers, and moles, and reptiles such as snakes and lizards also mix the soil and improve aeration and drainage, churning up rock fragments at the surface and exposing them to further disintegration. Their holes allow greater penetration of chemical processes especially during rainy seasons. Man himself has aided these weathering processes by removing minerals through the cultivation of his specialised crops.

Further work

See p. 27 for fieldwork exercises and questions.

4   Weathering and Climate

Weathering in the Humid Tropics

Chemical weathering is at its most active in the humid tropics where the heat and moisture of the equatorial rain forests encourage rapid growth of vegetation and its swift decay after death. Water passing through the rotting leaf litter washes down soluble salts which are then absorbed by plant roots to sustain growth. Some organic acids from the decaying vegetation combine with water in the regolith and, with high soil temperatures of 25°-28°C, chemical decay of rocks is rapid. Rock minerals are attacked chemically, quartz being the most resistant. Silicates such as micas and feldspars decompose to form new compounds—oxides of aluminium and iron, and clays. The aluminium and iron compounds combine with oxygen and water to create thick layers of laterite (p. 28) rich in iron compounds (ferricrete) or aluminium compounds (bauxite). These lateritic layers become rock hard or indurated to form duricrusts up to 10 m thick. Deep regoliths averaging depths of 30 m, but sometimes reaching depths of 125 m, are formed.

Under such intense chemical weathering rock resistance is important in the development of landscape. Crystalline rocks such as granites, gneisses and schists weather easily, providing quartz and feldspars to form the gritty clays typical of the rain forests. Gneisses and schists break down to form low hills separated by broad valleys and depressions (Fig 25). This forest-covered undulating landscape is occasionally broken by huge domed inselbergs of crystalline rock. These inselbergs suffer little mechanical erosion due to the low temperature range, and chemical weatherng on their surfaces is also reduced by rapid run-off. Such landforms may have originated during a drier climatic period.

Limestone offers little resistance to chemical weathering in the humid tropics. Thick, well-jointed beds of massive limestone above the water table in regions of moderate to heavy rainfall develop karstic landforms (p. 32). Below the water table the porous rock becomes saturated with warm acidic water and slowly dissolves to leave wide clay-filled marshy hollows at the surface.

Weathering in Arid and Semi-arid Regions

Deserts and semi-deserts are regions of low rainfall, high evaporation rates, and large daily and seasonal temperature ranges, and have a clear dry atmosphere allowing high absorption and radiation of solar energy. Little moisture is available for plant growth and large areas of open ground are exposed to the weathering process. Consequently, surface landforms are largely sculptured by mechanical forms of weathering which produce angular landscapes of inselbergs, escarpments and level erosion plains (pediplains, see p. 37); these are in contrast to the low

Fig 24.   Water movement and weathering processes in the tropical rain forest

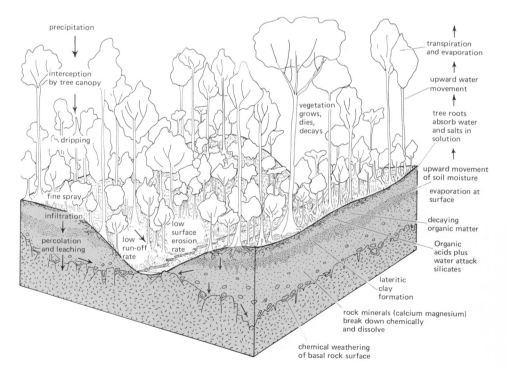

precipitation

interception by tree canopy

dripping

fine spray

infiltration

percolation and leaching

low run-off rate

low surface erosion rate

vegetation grows, dies, decays

transpiration and evaporation

upward water movement

tree roots absorb water and salts in solution

upward movement of soil moisture

evaporation at surface

decaying organic matter

Organic acids plus water attack silicates

lateritic clay formation

rock minerals (calcium magnesium) break down chemically and dissolve

chemical weathering of basal rock surface

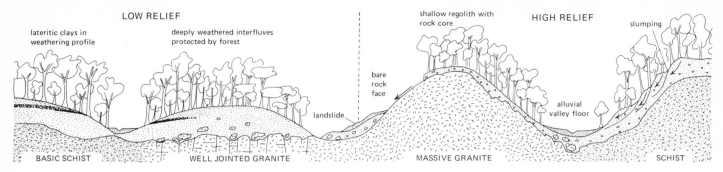

LOW RELIEF

HIGH RELIEF

lateritic clays in weathering profile

deeply weathered interfluves protected by forest

shallow regolith with rock core

slumping

bare rock face

landslide

alluvial valley floor

BASIC SCHIST · WELL JOINTED GRANITE · MASSIVE GRANITE · SCHIST

Fig 25. Landforms associated with different rock types in the humid tropics

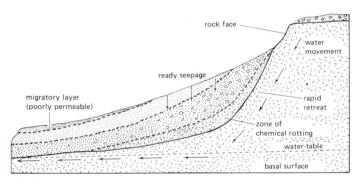

rock face

water movement

ready seepage

rapid retreat

migratory layer (poorly permeable)

zone of chemical rotting

water-table

basal surface

Fig 26. The weathering profile on a granite scarp in a humid tropical region. Chemical weathering causes the slope to retreat

rounded landforms of the humid tropics where chemical weathering and clay deposition tend to smooth out irregularities (Fig 28).

But although the dry atmosphere in arid regions reduces chemical and biological weathering these processes are still present. Chemical weathering can be seen on rock surfaces pitted with small holes called alveoles, sometimes excavated into small cavern-like forms termed 'tafoni' from which loose rock grains are removed by wind or water. On level surfaces chemical weathering attacks zones of mineral weakness to produce large shallow hollows called weathering pits.

Chemical action also produces desert varnish, a reddish-black mineral layer 2 mm to 7 cm thick which is well-developed on

Fig 27. Sierra Leone, central region: (a) domed inselberg rising above the tropical rain forest. Unlike the lower zones, run-off will be rapid from its surface and chemical weathering is at a minimum

volcanic rocks and sandstones. The varnish, which varies in age from 20 000 to 50 000 years, consists of an inner layer rich in silicates and an outer layer of manganese and ferrous minerals. It is probably caused by capillary action of chemically-rich water working itself to the surface. The varnish protects rock surfaces and slows down weathering.

The shady parts of desert rocks may also attract capillary moisture causing chemical rotting and here low forms of plant life such as lichens help retain some of the moisture.

But the large expanses of angular rock debris in arid regions are the result of intense heating and rapid cooling by occasional rain showers. Granular disintegration, flaking, exfoliation, pressure release sheeting, and rock splitting produce a coarse, loose, broken debris. Rough-textured igneous and metamorphic rocks such as gneisses and schists are eventually reduced to their individual grains; dolomitic limestone breaks down to dust particles; granite, easily affected by chemical changes, salt crystallisation and insolation weathering, is reduced to tiny quartz particles. Some types of sandstones are more resistant, however, since their grains are already the remains of weathering, resistant quartz being the chief mineral. Under similar climatic conditions, for example, the crystalline rocks of the Adrar massif in Mauritania have been weathered down to surfaces at 100 m to 200 m above sea level while nearby sandstones originally reaching the same height now stand at 400 m to 800 m. ·

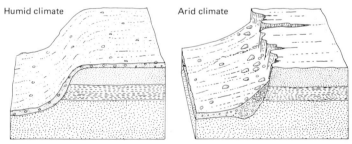

Humid climate

Arid climate

Fig 28. The difference in slope profiles in humid and arid regions

Weathering in the Savanna Regions

Between the humid tropics and the arid zones of Africa lies the savanna vegetation region of scattered deciduous trees and tall grasses experiencing alternate wet and dry seasons. During the rainy season, which decreases in length toward the desert fringes, the amount and force of the daily rainfall often equals that of the humid tropics. Chemical weathering is active on flat or gently sloping rock surfaces especially in the afternoon when the sun's heat is greatest. The rate of chemical weathering is, however, not as great as that in the rain forests because temperatures are never continuously high and, in fact, the onset of the rains often causes a swift drop in temperatures. Again, because run-off and percolation are rapid, especially at the beginning of the rainy season when vegetation cover is thinnest, rapid downward leaching of soluble minerals in the soils occurs, increasing the sands or clays at the surface. When leaching ceases during the dry season, the downward migration of minerals ceases. Bands of iron minerals (ferricrete), silica (silcrete), and calcium (calcrete) develop. Silcretes and calcretes are more typical of drier zones although fossil examples may occur in wetter regions. As in the humid tropics, duricrusts due to cementing by sesquioxides (Fe_2O_3 and Al_2O_3) develop to form distinctive landforms (Fig 29).

The onset of the dry season reduces chemical weathering although if the water-table remains high sub-surface weathering may still continue. Mechanical weathering—exfoliation and pressure-release sheeting—is probably most active at the beginning and near the end of the rains when temperature ranges are high and cooling is affected by frequent rain showers.

Weathering in the Warm Temperate Regions

During the cool rainy winters in the Mediterranean zones of Africa evaporation rates are lower and heavy frontal rainfalls cause much leaching or washing downwards of minerals in the soil. Chemical weathering continues but is only a fraction of the rate of the humid tropics due to lower temperatures. On the highest peaks of the Drakensberg, on Table Mountain and the higher parts of the Cape and Atlas ranges mechanical freeze-thaw weathering is active with periglacial landforms occurring between 1000 m and 1700 m. During the summer droughts, water-tables drop, chemical weathering is greatly reduced and mechanical weathering similar to that of the semi-arid regions predominates.

The Development of the Weathering Profile

The regolith is the upper zone of deeply weathered rock in which minerals and rock particles are transported and removed by water. Saprolite is rock weathered where it stands without any movement occurring—the rock rots *in situ*. The

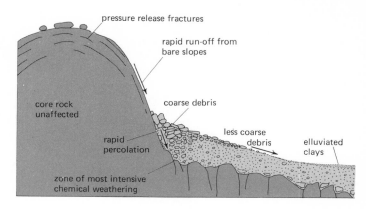

Fig 30. Weathering and erosion on a domed inselberg in the savanna landscape

weathering front or basal front is the boundary below ground between the weathered and unweathered rock zones. Regolith and saprolite are deepest on flat or gently undulating surfaces especially in the humid tropics and where there has been little climatic change over long periods. In Uganda 45 m is a common regolith depth and 50 m in the igneous and metamorphic rocks of the Jos and Kano plateaux of Nigeria. Between Abeokuta and Meko in southwest Nigeria, regolith depths vary from 5 m to 70 m.

Weathering of rock is much more active above the water-table than below it where water is moving less freely. The fewer cracks and joints at lower depths also make weathering penetration more difficult. Deep regoliths take a long time to develop: in parts of the Ivory Coast the granite has rotted to a depth of only 1 m over periods ranging from 22 000 to 77 000 years, while in Uganda it has taken about 1 million years to form a 9 m-deep regolith. Most mature regoliths in the tropics are over 20 000 years old.

The joint system of a rock may hasten rock rotting processes by allowing deep, sub-surface penetration of the weathering elements. Acidic rainwater penetrates along vertical and horizontal joints and rots the joint faces. The corners of sub-surface joint blocks are weathered more rapidly than the faces to produce sphere-shaped core-stones, a process termed spheroidal weathering. The degree of weathering of such core-stones is a good indicator of the extent of the weathering process.

Eventually, the regolith or saprolite will acquire several zones decreasing in weathering intensity with depth (Fig 31). The highest zone is weathered rock *in situ* and is followed by a zone of decomposed rock still retaining its structure. Below is a zone of rotted rock containing rounded core-stones, then a zone containing large angular core-stones overlying unaltered bedrock.

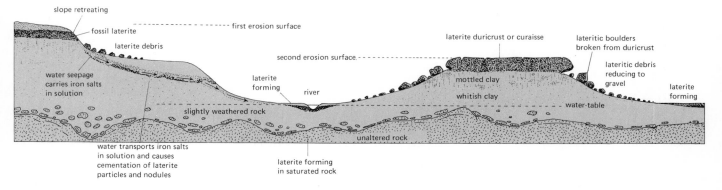

Fig 29. Laterite formation and mesa landforms in the savanna landscape

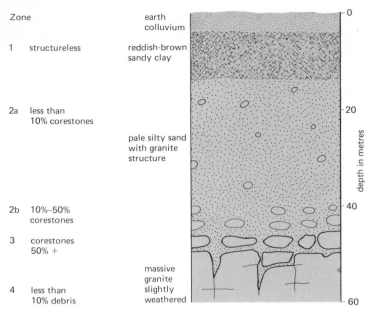

Zone		
1	structureless	earth colluvium
		reddish-brown sandy clay
2a	less than 10% corestones	pale silty sand with granite structure
2b	10%–50% corestones	
3	corestones 50% +	
4	less than 10% debris	massive granite slightly weathered

depth in metres: 0, 20, 40, 60

Fig 31. A regolith profile in the humid tropics. Note the great depth of rotted material

Denudation Processes

The level of the landscape surface is constantly being lowered by the removal of rock waste by flowing water—sheet wash, streams, and rivers. As the surface level is lowered the rotting process is able to penetrate more deeply into the unweathered rock below ground. This removal of surface waste is termed denudation and can be accomplished by chemical or mechanical means or a combination of both.

Chemical denudation occurs when groundwater seeps into the regolith and then flows laterally carrying dissolved salts into streams. Measurements show that rivers in tropical Africa carry an average of 10 parts of silica for every million parts of water, a figure slightly above that for rivers in temperate regions. This indicates that much of the silica released in the weathering

Fig 32. Zimbabwe, near Salisbury: spheroidally weathered boulders of granite. Note the honeycomb weathering, exfoliation scars and the joint structure

process goes to form clays or is absorbed by plant roots. In the cool rainy Eastern Highlands of Zimbabwe (1220 mm a year) tests show a rate of chemical loss by solution from granite of 400 kg per hectare in one year, a surface lowering of 15·4 mm every 1000 years. In drier years (920 mm a year) the surface lowering is 5·9 mm every 1000 years. Observations taken in the savanna regions of Africa show that dissolved salts represent between 34% and 38% of a river's load compared with up to 70% in the rain forests. This demonstrates the greater efficiency of chemical weathering in the warm, wet humid tropics.

Mechanical denudation involves the mass movement or wasting of regolith by soil creep, landslips, landslides, slumps, slope wash, and by streams. Soil creep is very active in the savannas due to periodic wetting and drying causing soil particles to expand and contract and move downslope under gravity. Overhanging river banks, low ridges on sloping ground, angled trees, boulders and large stones in river valleys are evidence of creep. Landslides are common on steep slopes but rare on the flat plateaux of Africa. They occur in deeply weathered rocks, clays which hold a lot of water, or on porous soils. The water acts as a lubricant along the junction between regolith and unweathered rock, sliding occurs and the unweathered rocks are uncovered. Exposed duricrusts allow water to penetrate through surface cracks and, if they overlie clay, the crust may become unstable and break up into blocks which slide during heavy rainstorms. Slumping is common in the humid tropics especially where saprolite is removed by washing out (eluviation) below a protective crust, e.g. during the wet season in the Sula Mountains region of interior Sierra Leone.

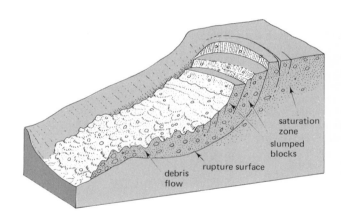

Fig 33. Type of slump caused by saturated regolith in a humid region

Slope wash—the transport of debris over slopes by broad sheets of water—is more common in the savanna regions especially at the beginning of the rainy season. The impact of a single raindrop can move a stone of 4 mm diameter some 20 cm down a slope and on steep surfaces stones of up to 400 mm diameter are moved. Slope wash also occurs in the equatorial rain forests where tree removal has exposed the ground surface to rain attack: in undisturbed rain forest about 1 kg per hectare is lost each year, 34 tonnes per hectare if the trees are cut down.

Deposition of Weathered Material

Weathered materials washed into depressions and basins form widespread mantles of fine materials—clays and sands—usually not more than 1 m or 2 m deep. In the drier savannas deposition is more widespread and the mantles are thicker and coarser. In

Fig 34. Nigeria, near Nanka, East Central State: severe gulleying and slumping is removing soil, regolith and vegetation

arid areas major landforms result from such deposition—alluvial fans, playas, deep sand sheets, and debris-filled hollows (pp. 107–8). In the wetter parts of the savanna, e.g. on the Jos Plateau and in the Congo/Zaïre basin, similar landforms may be traced to the dry interpluvial periods of the Pleistocene.

Fieldwork

Examine your local area for evidence of pan or crustal formation. What are the dominant weathering processes in your region? Examine rock faces for microfissures and colouring due to chemical weathering. Obtain scrapings from rock surfaces and analyse them in the laboratory. What differences are there between the coloured surface scrapings and the non-coloured? Explain. Chip a piece of weathered rock from the surface and a piece of freshly exposed rock. Compare the two. Start to collect small pieces of rock, analyse them in the laboratory, and classify them.

Questions

1. Under what climatic conditions in Africa does mechanical weathering dominate chemical weathering?

2. In which parts of Africa, and for what reasons, is chemical weathering a dominant process?

3. Discuss the influence of climate on the nature and rate of rock weathering.

4. What are the characteristic processes of weathering in (a) the humid tropics; (b) arid regions; (c) warm temperate (Mediterranean) regions; (d) savanna regions, of Africa?

5. What is the difference between regolith and saprolite? With reference to specific examples in Africa and using diagrams, describe the development of the weathering profile.

5   Weathering and Landforms

Laterite Forms

We have already seen how downward leaching of soluble minerals in the soil and mineral migration may result in the formation of mineral-rich layers (ferricrete, silcrete, calcrete). Such processes have continued over long geological periods so that some mineral layers have been fossilised into rock-hard bands. Erosion may later expose such 'crusts' to form distinctive landforms. The most widespread of these rock-like fossilised crusts in tropical Africa is termed laterite, a relic of an ancient phase of soil formation.

Laterite is composed of highly weathered rock minerals rich in iron, manganese and/or aluminium oxides which are concentrated in thick layers. Iron-rich laterites are sometimes called ferricretes but where heavy leaching has removed iron and silica, laterites rich in aluminium compounds (bauxite) are formed. Laterite contains few silicates but often has large amounts of kaolinite (a very fine clay) and quartz cemented together by sesquioxides. Chemical changes caused by wetting and drying and by heat cement laterite particles into rock-hard or indurated layers either on the surface or a few metres below.

Laterite layers are usually about 2 m thick but may reach 12 m. They are the result of the long and continuous weathering of rock: laterites containing between 30% and 80% iron oxides are known to have originated from granite or dolerite which contain from 11% to only 2% iron oxide, therefore the original rock must have been several times thicker than the present lateritic layer. The 9 m-thick laterite cappings in Buganda, for example, have been derived from rock 180 m thick.

Some of the iron compounds found in laterites have been produced organically in swamps or by capillary action in the regolith, but most are derived from rotted rocks and percolate downwards and laterally. The speed of this movement is greatest in uplifted regions, the particles migrating towards the deeply incised valleys of rejuvenated rivers (Fig 36).

Ideal conditions for laterite formation are found in regions of low relief with over 1000 mm of rain a year and where a dry season allows the iron compounds to lodge fairly near the ground surface. Such conditions may occur throughout the savanna regions, on the desert margins and in the humid tropics, although the length of the dry season varies considerably.

Indurated laterites form distinctive landforms in the tropics and semi-arid zones of Africa. The tableland or mesa is common varying from small flat-topped hills to extensive

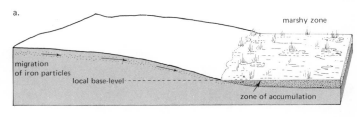

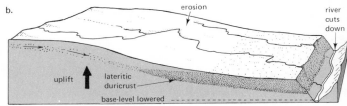

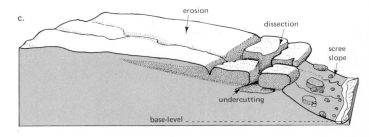

Fig 36. Laterite formation and dissection: (a) Iron particles migrate in seeping water, accumulate in swamp. (b) Land uplifted, swamp drained by rejuvenated river, laterite dries and hardens as duricrust. (c) Lateritic cap slowly dissected by surface streams and percolation

Fig 35.   Nigeria, Jos Plateau: a laterite mesa with granite corestones in the foreground

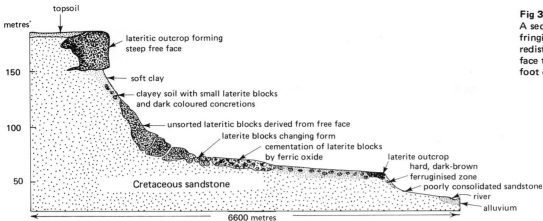

plateaux. These mesas end in steep laterite cliffs a few metres high overlooking gentler slopes strewn with large boulders of break-away laterite (Fig 37). Resistant laterite bands jut out like terraces on some slopes. Laterite fragments fallen from cappings are sometimes broken down and re-cemented into new layers called pavements or provide infilling for shallow surface hollows.

Laterite forms may be eroded in several ways. Water seeping through fissures in the laterite cap emerges in springs at the base of the cliffs causing them to crumble and collapse. Caverns in the regolith undermine the laterite layer and gradually the mesa slopes retreat.

Laterite landforms are prominent in the Sula Mountains and Gbenga Hills of Sierra Leone, in the aluminous laterites of Ghana, on the Jos and western plateaux of Nigeria, and in the Buganda Province of Uganda.

Fig 38. Tanzania, near Mwanza, Sukumaland: the surface of an exposed granitic batholith has been weathered into tors and spheroidal boulders. The boulders in the foreground are slowly emerging from lateritic material which is being lowered by surface wash

Domed and Boulder Inselbergs

The word inselberg (German—island mountain) describes any isolated hill or hills which stand prominently above a level surface. Inselbergs include laterite-capped masses of saprolite, a hill of sedimentary rock, a castle koppie, a tor of rounded core-stones, or a massive rock dome with vertical sides called a bornhardt or domed inselberg. Such landforms are often seen together in the landscape and those formed from plutonic crystalline rocks are very common. Although some domed inselbergs occur in massive unjointed rocks, most tors and castle koppies are formed in strongly jointed rock, usually granite, the joints being caused by stress during cooling and crystallisation, by crustal movements, or by pressure release.

Tors are piles or ridges of spheroidally weathered boulders often of granite which have their bases in the bedrock and are surrounded by weathered debris. They vary in height from 20 m to 35 m and have core-stone diameters from 3 m to 8 m. They may occur at high points in the landscape (skyline tors) or in depressions and on hill-sides (sub-skyline tors). Rock piles which have suffered little spheroidal weathering appear cube-like and blocky and are known as *castle koppies*.

Tors are due to the chemical rotting of rock along joint

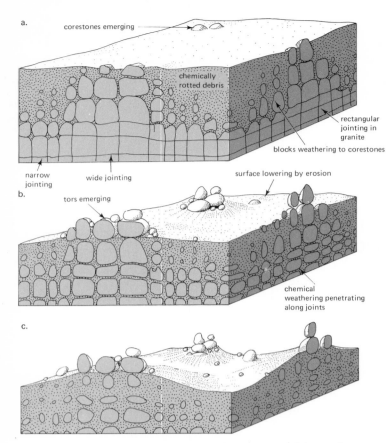

Fig 39. Block diagrams showing the development and decay of granitic tors in a savanna landscape

planes below the surface. If the joints are widely spaced the core-stones form huge blocks, but if the joints are close smaller stones result which are easily weathered and removed from between the bigger blocks (Fig 39). If the denudation rate of the regolith surface exceeds the chemical weathering rate of the blocks below the surface at the weathering front, the blocks will eventually be exposed above the ground level. Gradually the rocks below the surface which support the tor will be chemically rotted and cause its collapse.

Fig 40. Tanzania, near Masasi: a domed inselberg or bornhardt in the 'inselberg corridor' of southern Tanzania

Tors and castle koppies are familiar landforms in many parts of Africa. The Mwanza district of Sukumaland in Tanzania has many examples, a well-known one being Bismarck Rock which stands in the shallow offshore waters of L. Victoria. In Uganda granite tors rise to 130 m above nearby valley floors in the Mubende region. Tors also overlook the undulating plains of the Jos Plateau in Nigeria, e.g., between Kaduna and Zaria. Castle koppies, tors and bornhardts are common in the Matopos region near Bulawayo, and around Salisbury, Marandellas and Macheke in Zimbabwe.

The monolithic domed inselberg or *bornhardt* is a characteristic landform of the granitic plateaux of the African savanna but also occurs in the semi-arid and humid tropical regions. Some domed inselbergs rise almost vertically from pediplains (p. 37) to rounded summits 300 m high. Others are lower, up to 100 m, and appear as huge swellings of granitic rock called whalebacks or dwalas (in Central Africa). Low rock pavements which hardly break the surface but which have steep vertical sides below ground (ruwares in West Africa) may later emerge as bornhardts. Some domed inselbergs have a less steep rock slope or pediment at their base, others form a sharp angle between the plain and the bornhardt sides. Some bornhardts are symmetrical, others asymmetrical. The upper surface is usually convex but may also flatten out and have small lakes in weathered summit hollows.

Bornhardts develop only in plutonic igneous or metamorphic rocks such as granites. These rocks develop curved joints due to pressure release and this jointing, occurring up to 35 m below the rock surface, controls the 'shelling' of curved rock sheets. Vertical jointing is also present in granites and leads to the breakdown of domed inselbergs into castle koppies (Fig 41).

There are two major theories which explain the formation of the domed inselberg. The stripping or exhumation theory entails the increased removal of the regolith so that the unweathered rocks below the surface are exposed as bornhardts. In contrast, Lester King's scarp retreat theory involves the parallel retreat of valley sides or lateral scarps until only remnant inselbergs are left. Both these theories are discussed more fully on pp. 37–40.

Although bornhardts are very resistant to weathering they can still be attacked. The upper steep slopes suffer very little weathering for here run-off is rapid and chemical weathering is therefore small. But intense chemical weathering occurs along the bornhardt base where rainwater collects. Here silt and clay are being removed in solution and by mechanical wash (eluviation)—processes which eat into the foot of the inselberg. Such

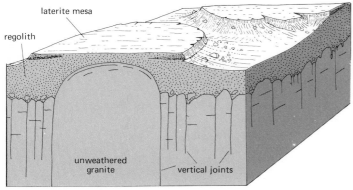

a. Laterite mesa landscape

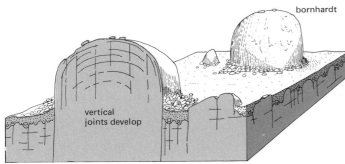

b. Bornhardts begin to emerge

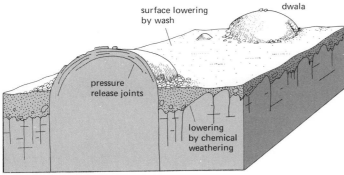

c. Bornhardts dominate plains

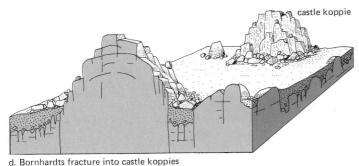

d. Bornhardts fracture into castle koppies

Fig 41. Block diagrams showing the development and decay of the domed inselberg or bornhardt in a savanna landscape

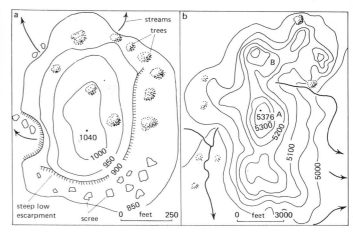

Fig 42. Contour sketch-maps of two domed inselbergs: (a) Aseke Hill, near Oyo, Western State, Nigeria. (b) The twin domed inselbergs of Dombashawa (A) and Mashawamuura (B), 15 km north of Salisbury, Zimbabwe

basal sapping undercuts the inselberg and causes sheets of rock to fall from the sides. So the bornhardt is very slowly reduced in circumference.

Domed inselbergs are a common feature of the southern 'inselberg corridor' of Tanzania where striking examples rise abruptly from the level plains to heights of 300 m. Classic examples occur in Zimbabwe, particularly in the Matopas area southwest of Bulawayo, at Domboshawa (Fig 42) north of Salisbury, and in the northeastern part of the country. Bornhardts are a familiar feature in Nigeria especially around Bauchi and Zaria and in Ekiti District (the Amoye inselbergs) of Western Province. Individual outstanding examples are Zuma Rock in Niger Province, Abuja Mt in Northern Nigeria, and Aseke Hill (Fig 42) in the banded gneiss of the Oyo District, Western State. A superb example in Malaŵi is Hora Mt in the Mzimba District.

Etchplains

Gradual lowering of the regolith surface by stripping may also explain the extensive exposed rock plains studded with inselbergs called etchplains which occur in the tropical and arid regions of Africa. Etchplain formation is explained by J. Büdel (1957) who states that there are two levels of weathering in the African plains—one at the surface and one at some depth below the surface. He maintains that rivers in tropical Africa are not deeply incised into regolith surfaces but flow over them, constantly meandering and changing direction and flooding during rainy seasons. The rivers wash away the finer particles of the regolith, gradually lowering the surface. At some depth below this regolith surface or wash surface lies the weathering front where chemical processes are attacking unweathered rock. The whole of the savanna plain is thus being lowered by wash at the surface and by chemical decay at the weathering front below the surface (Fig 43). If the upper surface is lowered at a rate faster than at the weathering front, then the latter surface will begin to emerge as an etchplain. Such etchplains are common in the deserts of Africa, for example, on the Sahara plains, in the Hoggar and Adrar des Iforas highlands, and in southern Somalia.

Etchplains may be seen at various stages of development

(Fig 44) and, using examples from West Africa, F. J. Thomas (1965) has outlined these stages as:

(a) *Laterised etchplain*: low relief, extensive laterite deposits, little stream incision, few residual hills.

(b) *Dissected etchplain*: swifter stream erosion, duricrust breaking up, valleys more deeply cut, mesas. Some exposure of the weathering front as tors and domes.

(c) *Partly stripped etchplain*: further dissection and waste removal. Most laterite now removed. A few mesas, tors, dwalas, ruwares, castle koppies, bornhardts.

(d) *Stripped etchplain*: much rock exposed as convex hills, some basins still containing sediments.

(e) *Incised etchsurface*: the weathering surface is now exposed and is being eroded on a large scale by rivers.

There is still much research to be done on etchplanation which is obviously an important process in the development of landforms in arid and tropical regions.

Karst

Limestone (calcium carbonate, $CaCO_3$) and dolomite (calcium and magnesium carbonate, $CaCO_3MgCO_3$) are readily soluble in rainwater (carbonic acid, H_2CO_3) and as a result limestone or dolomite occurring in regions of moderate to heavy rainfall develop distinctive landforms. The term 'karst', named after a limestone region in Yugoslavia, was first used by J. Cvijic (1918) for characteristic limestone landforms and his description of them formed the basis for investigation elsewhere. Research in the humid tropics, however, has revealed karstic landforms very different from those of Cvijic's temperate karst.

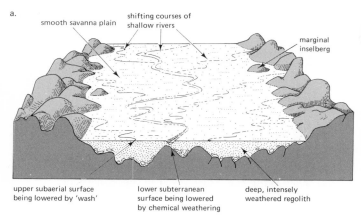

Fig 43. The formation of the plain and inselberg landscape according to J. Budel. (a) The regolith surface is being lowered at a faster rate than the subterranean surface. (b) The irregular subterranean surface is being exposed as inselbergs.

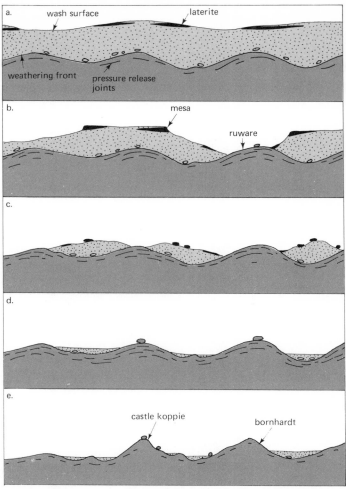

Fig 44. The development of etchplains and etch surfaces according to M.F. Thomas. (For explanation see text.)

The Karst Cycle

Cvijic's karst cycle still provides a basis for understanding karst formation as a whole and may be summarised as follows:

(a) *Youth*: Rainwater contacts exposed limestone causing dissolution along lines of weakness. The surface is etched into lapiés or clints and grikes—narrow vertical channels separated by sharp narrow ridges. A red clay residue termed *terra rossa* may fill the grooves. Joints and crevices are enlarged by solution and dolines (solution pans) develop into shallow funnel-shaped depressions. These sometimes become clogged with inwash clay and hold water to create a karst lake. There is little underground drainage and no large caverns.

(b) *Maturity*: Surface drainage now flows into enlarged openings called swallow holes and disappears underground via sinkholes to join a developing subterranean drainage system. Surface streams are few and flow for only short distances, usually on clay-lined channels. Water descends along bedding and joint planes, widening them by solution and erosion to create an extensive cavern network.

(c) *Late maturity*: Extensive subterranean solution increases structural weakness causing surface collapse. Underground streams become visible through surface karst windows, and solution valleys—long narrow irregular depressions—develop. Several sinkholes may merge to form compound sinkholes and extensive depressions called uvulas, with natural bridges, are formed by roof collapse. Poljes—elongated basins with flat floors and steep enclosing walls—are created by fault-controlled erosion and may contain large lakes.

(d) *Old age*: Remnant inselberg-like hills termed buttes, temoines, hums, mogotes or haystack hills, now dot the landscape and are separated by broad, flat basins covered with clay residues. The hums may be riddled with sinkholes and cavern networks.

Certain conditions are necessary for such karst formation. The soluble limestone or dolerite must be highly jointed rather than permeable throughout; karst landforms are poorly developed in chalk where rainwater soaks through *en masse*. The subsurface water must not remain stationary, but should drain away to deep valleys outside the region, for karst features will not form where water collects in the rock.

There is not yet full agreement on how caverns are formed in karst regions. Some geomorphologists believe caverns are formed by solution below the water-table and that when the water-table is lowered the old channel is abandoned and work begins on the formation of a new cavern network. Others consider that caverns form at the water-table level rather than below it and if the water-table falls the process begins again along the new lower water-table level (Fig 46).

Most limestone regions have underground caverns at various levels, each level connected by conduits or narrow passages caused by solution (Fig 47). Such elongated or galleried caverns develop along bedding planes enlarged by solution. The walls, roofs and floors of the caverns are usually covered with cave travertine or tufa, features caused by calcium carbonate deposition from water seeping through to the cavern roof. The calcium bicarbonate in the water changes back to calcium carbonate on contact with air in the cavern and, as water drips from the roof, tiny particles of calcium carbonate are left behind in the form of icicle-like columns called stalactites which slowly grow by the addition of more particles. Drops splashing on the floor also leave calcium accretions which build up into stalagmites. Stalactite growth is governed by gravity and a central feeder pipe, but stalagmites have no structural control and are mound-like. Some hanging forms called helictites grow irregularly due to the orientation of individual crystals. Irregular shapes caused by flowing water on the cavern floor are termed flowstones. Sometimes the floor is masked by clay and silt and interconnecting conduits may become blocked with such debris.

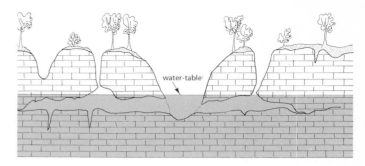

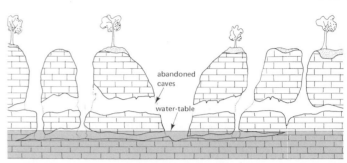

Fig 46. One theory of cavern formation in a limestone region: (a) Caverns form by dissolution of limestone at, or just below, water level. (b) Water-table drops, old caverns abandoned, new caverns forming at new water level

Many of the karst features described above exist in various parts of southern Africa. The dolomite plain of the southwestern Transvaal displays the monotonous flat surface of approaching old age with irregular surface drainage and an underground cavern and drainage network. Numerous sinkholes and dry valleys indicate the lowering of the water-table and rivers such as the Molopo, Notwani, Klip and Hartz issue from rises or springs (locally called eyes) at the head of blind valleys. Several solution valleys occur near Lichtenburg and large caves up to 2·5 km long exist at Sterkfontein, Makapansgat and Randfontein (Fig 52). Dolines and sinkholes also occur in the southern Transvaal. In the Carltonville-West Driefontein gold-

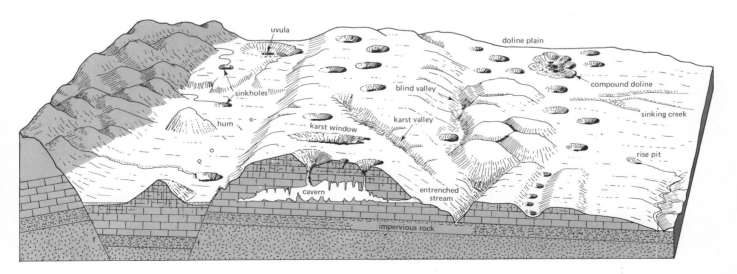

Fig 45. Major landforms associated with karst landscapes

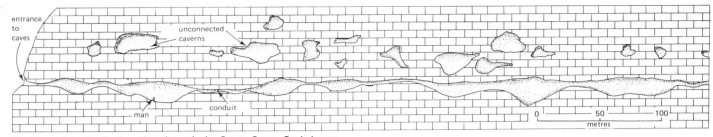

Fig 47. Longitudinal section through the Cango Caves, Oudtshoorn, southern Cape Province, South Africa. The level line of the lower caves suggests they were formed along the water-table

mining area, sediment-filled depressions and sinkholes in dolomite beds of the Jurassic period have been discovered. Karst features are also located in the Kaap Plateau, at Taungs in the northern Cape Province, in the Otavi district of Namibia, and near Oudtshoorn where the extensive Cango Caves are a major tourist attraction.

In Zimbabwe, several caves occur in young limestones near Sinoia west of Salisbury and there are ten sinkholes of varying size and depth. The caves contain a subterranean lake of a deep blue colour due to dispersed lime particles. Karst features on a small scale occur also on the old coastal plain inland from the coasts of Kenya and Tanzania and on the *uwanda* (coral platform) of Zanzibar and Pemba islands. In the northern Middle Atlas the action of melted snow has produced *nivo-karst* landforms in the high limestone mountains.

Karstic Landforms in the Humid Tropics

In the humid tropics distinctly different karstic landforms emerge. Compared with temperate regions, rainfall amounts and intensity and the activity of biological carbon dioxide (CO_2) are much higher. Greater evaporation rates cause ground surface deposition of calcium carbonate as travertine, and stalactites overhang cliff faces and cave openings. Within the caverns temperatures exceed 21°C and relative humidities are lower than in temperate regions. Excessive stalagmite formation occurs as well as case hardening—the formation of thin hard layers of calcium carbonate over deposition surfaces.

Mechanical erosion by flowing water is more active in the humid tropics than in temperate regions. Intense heavy showers

Fig 48. South Africa, Southern Cape Province: stalactites, stalagmites and columns of calcium carbonate in the Cango Caves near Oudtshoorn

cause rapid run-off and surface streams flow for some distance before they plunge underground. Joints are rapidly widened and surface debris is washed below and transported to areas outside the region. A landscape emerges of steep hills with rounded or conical tops enclosing irregularly shaped depressions (contrast the low hills and shallow depressions of the temperate karst).

Variations in climate and rock structure are responsible for the following variety of karstic landforms in the humid tropics:

(a) *Cockpit, cone or kegel karst*: Hills are cone-shaped (slopes 20°–40°) and enclose circular or elongated ravine-like depressions. Star-shaped depressions or cockpits are formed by solution along major joint intersections. The depressions eventually become flat alluvium-covered plains overlooked by steeply rising conical hills (Fig 51).

(b) *Tower or turm karst*: The hills are steeper (60°–90° slopes), the massive, consolidated limestone producing cliff faces and jagged summits. In later stages karst towers, riddled with cavern networks, stand up as massive pinnacles above alluvial plains (Fig 51). Small shallow caves called *foot caves* form at the base of the towers.

(c) *Crevice karst*: Here the surface is criss-crossed by enlarged joints up to 6 m wide and 20 m deep. After long periods of weathering the main joints become deep chasms bounded by vertical rocky faces. Adjoining faces retreat to create a landscape of jagged ridges and pinnacles termed *arête and pinnacle karst*.

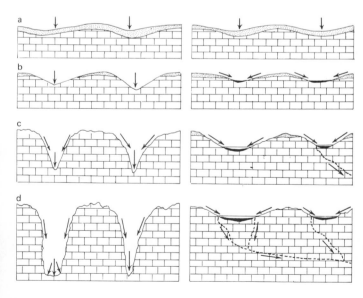

Fig 49. Denudation of karst: *left:* in a humid tropical region with more than 3 000mm of rainfall per annum; *right:* in a temperate region with less than 700mm of rainfall. The arrows represent the direction of water flow

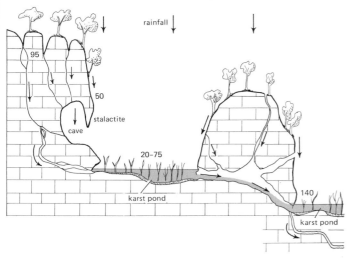

Fig 50. The rate of limestone dissolution in a region of humid tropical karst. The figures indicate the approximate number of milligrams of dissolved limestone in each litre of water sampled at different points

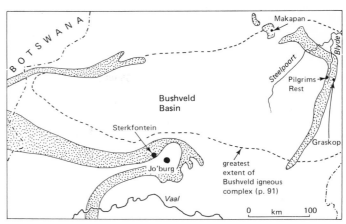

Fig 52. Outcrops of dolomitic limestone in the Transvaal Province, South Africa

Much of the research on tropical karst has been in southeast Asia, the Caribbean, and Central America with rather less attention being given to the regions in Africa. Here the greatest extent of humid tropical karst landforms are located in the limestones of the Kahavo-Ankara area and the Mahafaly Plateau of the central and southern parts of the low plateaux of western Madagascar, and in parts of the Cameroon, Equatorial Guinea and Gabon republics. The karstic landscape of the Eastern Transvaal dolomites in the Blyde Canyon region also presents an unusual surface (for a sub-tropical region) of conical hills and wide flat-floored depressions. The region is one of high seasonal rainfall (1000 mm annually between October and April) and the relief is influenced by the perennial Blyde River and its tributaries. The Blyde has cut a canyon controlled by fault lines nearly 1000 m deep into the dolomite plateau. Numerous dykes, sills and bands of chert, quartzite and shale intrude into the dolomite and often outcrop on the slopes of several steeply rising conical hills (Fig 53). These hills appear to be a form of kegel karst of the tropical humid type. Some are low rounded hills rising about 50 m above the

plateau surface, others reach to 200 m above the Blyde Canyon. They appear less steep now than formerly and contain several caves well above the present water-table. The hill surfaces are marked by sinkholes and lapiés. The presence of these conical hills and their associated basins masked by deep alluvial beds similar to tropical karst plains, suggests formation under warmer, moister tropical conditions, possibly during the Pleistocene period.

Fieldwork

(Note: The fieldwork can be attempted only by students living in appropriate regions.)

1. Laterite: (a) Examine your local area for evidence of laterite formation, e.g. on unmetalled roads and tracks or in cleared forest. Is the laterite ferruginous or aluminous? How hard is the surface? Can it be broken with a hammer only or is it relatively soft? Is it a surface layer or a pan formation at some depth? If possible, measure the thickness of the layers and the depth at which they lie. What are the climatic characteristics of the region? Obtain a sample from the cleared surface and a sample from the same area under forest. Both samples should fill tins of the same size and be weighed. Is there any weight difference? Examine for differences in biota content and break up the samples and examine the individual grains under a magnifying glass. What are the colour differences?

(b) Have any particular landforms been caused due to laterisation in your area, e.g. mesas? Sketch these landforms, measure their slope angles (p. 44), note any prominences on the slopes, measure the heights of the features. Examine the surface for fissures. How does water drain away? Are there springs at the base of the mesa and are they sapping the slopes? Is there evidence of re-cementation of laterite around the mesa? Write up your findings and explain them.

2. Tors: Are the tors in your region of the spheroidal type or angular castle koppies or a mixture of both? Plot their distribution in a measured area and draw a landscape sketch to show the general aspect. Are the tors skyline or sub-skyline? Plot these differences. Examine one or two typical piles. What is the height? What are the diameters of the exposed core-stones? Is there any uniformity in the size of the upper boulders in the district and in the lower boulders? What stage have the tors reached; are they old and collapsed or are they upright? Examine the under parts of the lower boulders for any signs of chemical erosion. Examine the scree surrounding the tors and measure its thickness and the sizes of individual pieces.

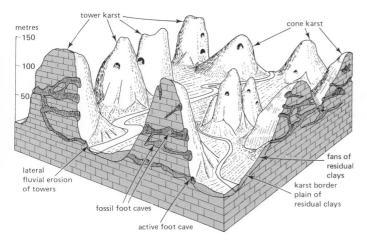

Fig 51. Block diagram showing features associated with a tower and cone karst landscape in a humid tropical climatic region

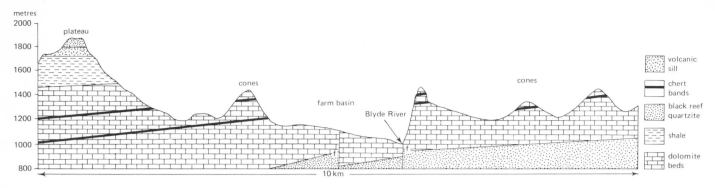

Fig 53. A section across the Farm basin showing the geology and the surface cones. Locate this area on the map, page 129

Legend:
- volcanic sill
- chert bands
- black reef quartzite
- shale
- dolomite beds

What is the depth of exfoliation scars and pressure release scars? Of what type of rock are the tors composed?
3. Domed inselbergs: What is the angle of slope at the base of an inselberg and higher up its sides? Draw a field-sketch. Pace or, if possible, measure the distance round the base. Is the plan round or oval? At what angle does the base plain slope away from the inselberg? Is there evidence of basal sapping? Are the inselbergs in the region of the 'whale-back', 'turtle-back', or conical type?

Questions

1. What is laterite? How is it formed? What landforms are due to the presence of laterite?

2. Define the term 'inselberg'. Explain, with the aid of examples, the difference between tors, castle koppies, and domed inselbergs (bornhardts).

3. What is an etchplain? Describe the development of etchplains with reference to examples in Africa.

4. Describe the differences in the morphology of karstic landscapes developed under temperate and humid tropical climates.

5. Choose three of the following landforms and describe, with the aid of examples and explanatory diagrams, their morphology (shape and appearance) and their mode of formation:

laterite mesa; tor; domed inselberg; etchplain.

6. Examine the development of landforms in jointed crystalline rocks under tropical climatic conditions.

Fig 54. South Africa, the Blyde River Canyon, Eastern Transvaal: cone karst landforms are seen in background (right). On the left a stepped escarpment caused by rocks of varying resistance (volcanic sills, chert bands and quartzite). In the foreground the jointed dolomite has been extensively weathered into lapies (clints and grikes).

6 The Landscape Cycle: Slope Retreat and Planation

All landmasses which stand above sea-level are subject to weathering and denudation. Although no appreciable alteration of the landscape may be noticeable during a person's lifetime, slow changes due to denudation processes are continuing. Over a long geological period the rate of landscape lowering may be balanced by isostatic uplift but, if no uplift occurs, then in theory the land surface will be worn down almost to sea-level, the *base-level* below which erosion cannot normally take place.

Cycle of Erosion

The American, W. M. Davis, used the concept of base-level to explain the development of landscapes and was the first to suggest that the erosion of the landscape progressed in phases or cycles. Davis believed that the landscapes which had developed in the humid climates of his home region, Massachusetts, and elsewhere in North America, and Europe were the norm and that they had undergone a normal cycle of erosion passing through stages of youth, maturity and old age (Fig 55).

(a) *Youth*: An erosion cycle begins after the swift uplift of a landmass above sea-level. Rivers begin to erode this initial surface draining directly to the sea as consequent streams and attempting to erode down to base-level. Interfluves are broad, tributaries few, valleys deep, narrow and V-shaped.

(b) *Maturity*: The consequent streams are joined by subsequent streams developing along lines of weakness and flowing generally at an angle to the consequent streams. Subsequent streams are in turn joined by obsequent streams which flow generally opposite to the consequents. An extensive drainage network develops eroding the landscape surface. Interfluves are narrowed, the initial surface disappears, valleys broaden, meanders and floodplains develop. Many streams are now at grade, that is, their gradients are sufficiently steep to allow water to flow without much erosion.

(c) *Old age*: Little upland remains except for an occasional monadnock (Davis's name for an inselberg), flood plains are broad and stream gradients very gentle. Rock surfaces are buried beneath a thick mantle of waste. Very wide meanders, ox-bow lakes and cut-offs develop, tributaries are few and drainage sluggish. The monadnocks stand abruptly above a peneplain, their slopes gradually declining with vertical erosion.

These ideas of Davis captured the imagination of many students of geomorphology and for some sixty years landscapes were described in terms of youth, maturity and old age. But Davis's landscape cycle theory was based on his studies of humid regions; landforms and landscapes in arid, semi-arid, tropical and glaciated regions were attributed by him to climatic accidents which had interrupted the normal cycle. Outside these regions Davis believed that landscape-forming processes were similar in all regions. Researchers in tropical and arid regions, however, could not accept that 'climatic accidents' were responsible for the landforms so very different from those of the humid temperate zones. Moreover, vegetation cover, which Davis had neglected, appeared to have an important influence on landscape development. Again, vast level plains similar to Davis's peneplain were also part of the arid and tropical landscapes.

In the 1920s, a friend and contemporary of Davis, the German, Walther Penck, criticised Davis's ideas on erosion and his whole cycle concept. Slopes do not decrease in angle and become more gentle with age, stated Penck, and, indeed, there was no real evidence that landscapes experienced the stages described by Davis. Penck believed that slopes were eroded and retreated laterally and maintained their original angle—they retreated parallel to themselves. This is the basic idea of the theory of pediplanation.

a. Youth

initial surface

sea-level

b. Maturity

c. Old Age

monadnock

peneplain

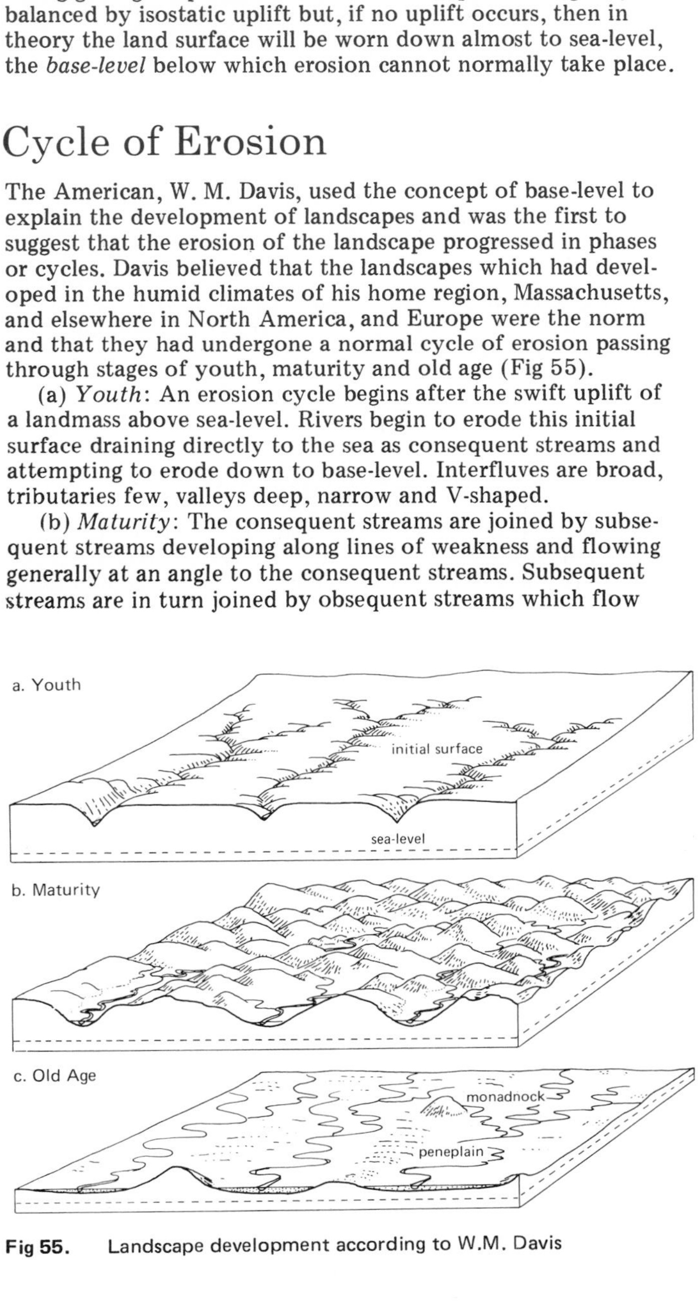

Fig 55. Landscape development according to W.M. Davis

The Theory of Pediplanation and the Parallel Retreat of Slopes

In the late 1940s, the South African geologist, Lester C. King, adopted and modified Penck's ideas and applied them to the landscape of Africa. His theory of pediplanation involves a close study of the nature of slopes, their erosion, and their parallel retreat. Four major elements of hill slopes are studied (Fig 56). The waxing slope or crest is the surface of the rock mass above the near vertical bare rock of the free face or scarp. Weathering causes the free face to lose material to the constant or debris slope below; this is cut into solid rock but may be covered by

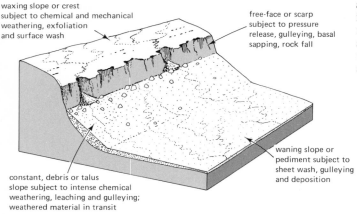

waxing slope or crest subject to chemical and mechanical weathering, exfoliation and surface wash

free-face or scarp subject to pressure release, gulleying, basal sapping, rock fall

waning slope or pediment subject to sheet wash, gulleying and deposition

constant, debris or talus slope subject to intense chemical weathering, leaching and gulleying; weathered material in transit

Fig 56. The four elements of study in the retreat of scarps

scree. From the base of the constant slope, the waning slope or pediment with a slope angle ranging from 1° to 7°, may stretch for hundreds of kilometres, its surface usually covered by assorted debris (Fig 56).

Weathering causes the free face to retreat slowly parallel to itself, the rock waste accumulating on the debris slope. The waxing slope becomes progressively smaller as the pediment develops until finally adjacent pediments merge to form extensive areas of gently sloping surfaces termed pediplains. Eventually the original land surface is reduced to isolated inselbergs which are finally eroded vertically and theoretically should disappear entirely (contrast this with Büdel's theory of lowering plains producing the plain and inselberg landscape, p. 31).

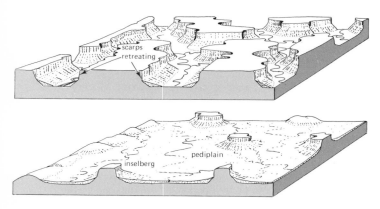

scarps retreating

inselberg

pediplain

Fig 57. Landscape development according to W. Penck and L.C. King. Lateral retreat rather than vertical lowering is the main process

The pediplanation sequence thus begins with a youthful stage of deep incision by streams, reaches maturity when approximately half the landscape consists of broad pediplains, and attains old age when the pediplains dominate the landscape Throughout these stages the escarpments maintain their original angles and retreat parallel to themselves.

There is considerable evidence to support this pediplanation theory. The American, A. N. Strahler (1950), showed that in small areas with the same climate, rocks, vegetation, soils and relief, slopes will retreat at a constant maximum angle and, if nothing disturbs this process, the landscape is in a state of equilibrium. Further research indicates that the presence or

absence of detritus or debris at the base of the slope is an important control of the retreat of the slope. If the debris is removed, and the presence of a river and tributary streams seems to be necessary for this, then the slopes retreat and maintain their characteristic angle. If the debris is not removed the slope becomes stationary and the slope angle declines (Fig 58).

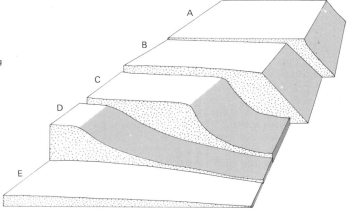

A
B
C
D
E

Fig 58. In the final stages of pediplanation vertical erosion exceeds lateral erosion thus removing the scarp

There is as yet not fully acceptable answer to the problem of pediment development and much more research is required on slope retreat in order to find solutions.

Climatic Geomorphology

The landscape cycle as imagined by Davis cannot be convincingly supported in Africa and has been rejected by the French geomorphologists, J. Tricart and A. Cailleux. Their research in the semi-arid and savanna regions of West Africa led them to criticise Davis because he stressed erosion and neglected deposition. They believe that landscapes can develop over much shorter periods than the millions of years which Davis thought necessary. Davis, they feel, tended to overstress fluvial action but neglected the effects of climate acting through the vegetation cover on landscapes.

To show how landscapes develop differently under different climatic conditions, Tricart and Cailleux refer to the Futa Djallon and the Mauritanian Adrar plateaux, both of which have reached Davis's 'stage of maturity' and have similar rock structures of thick sandstone but have different climates. The Futa Djallon plateau experiences drought from November to April and rainfall of from 1500 mm to 2000 mm from May to October while the Adrar massif has an arid climate with 100 mm to 200 mm rainfall annually. Each region has developed a particular landscape surface. The Futa Djallon is a region of deeply dissected plateaux with steep bounding escarpments surrounded by a wide zone of gentle surfaces (pediplains) dotted with sandstone inselbergs; major rivers have not incised their courses. The Mauritanian Adrar possesses a highly irregular surface drainage and the landscape is one of prominent westward-facing scarps, deep gulleys, and gentle eastward-dipping slopes. Here structure and climate have produced two entirely different landscapes. According to Tricart and Cailleux, pediplanation is a more valid theory in Africa than peneplanation.

The Concept of Dynamic Equilibrium

The studies of Strahler and others have shown that a condition of equilibrium or balance can be achieved during landscape development. In large regions where the controls of climate, rock structure and vegetation are uniform and have changed little over long periods, the processes of erosion, transport and deposition should strike a balance in the landscape. Drainage densities, slopes, and altitudes of hill and ridge crests will be similar throughout the region. Such conditions are seen in many part of Africa. This condition of equilibrium will continue so long as there is no interruption. Such interruptions as the land rising (isostatic movement), the sea-level rising or falling (eustatic change) or a change in climate are all events which could upset the landscape-forming processes.

Such a state of imbalance or disequilibrium would, however, gradually give way to one of equilibrium as the landscape and the processes acting upon it slowly adjust to the new conditions. Even over relatively short periods of time, climate can change from pluvial conditions (p. 111) to arid, from temperate to periglacial (p. 75), and from periglacial to glacial and the state of equilibrium will be temporarily disrupted. Then gradually a balance is restored. For example, slopes formed during cooler pluvial periods may have a convex form but as the climate becomes drier, mechanical erosion would increase, slopes would become steeper and the landscape more angular as it adjusts to new conditions of weathering and denudation. A new state of equilibrium involving new processes will be reached and will continue until once more disturbed by climatic change.

Many geomorphologists have adopted the dynamic equilibrium approach to the study of landform and landscape. Landforms are seen not as constantly developing and changing but as maintaining their forms. Slopes retreat, therefore, but they retain their characteristic angles. The attention of the student of landscape is diverted from the study of theoretical stages of past development to the study of the landscape as it is now.

Further Work

See page 44 for fieldwork exercises and questions.

7   The Erosion Surfaces of Africa

If we accept the ideas of King and Büdel then the characteristic plain and inselberg landscape of Africa is due either to scarp retreat and pediplanation or deep vertical weathering creating plains, emerging inselbergs, and retreating fringing scarps.

The plain and inselberg landscape is sometimes called the savanna landscape and examples of it are numerous: the Masai Steppe, the central plateau and inselberg corridor of Tanzania; the Mzimba and Lilongwe plains of Malaŵi; the Sabi Valley and the Gwelo-Salisbury high veld of Zimbabwe; the Great Karoo and the Springbok Flats of the Central Transvaal in South Africa; the Namib and Namaqualand regions of Namibia; the Accra interior plains especially in the Krobo and Shai districts of Ghana; the Wute Plain of Central Cameroon; the plateaux of the Oyo region and Hausaland in Nigeria; the Hombori region of Mali as well as the pediplains of the Sahara and Libyan deserts (Chapter 14).

The monotonous level pediplain forms extensive landscapes in southern and central Africa where L. C. King conducted most of his research. King rejected the peneplanation cycle of Davis as a suitable explanation for these vast erosion surfaces separated by steep escarpments. These, he argues are due to the process of parallel scarp retreat from incised rivers which have created extensive pediments. Over millions of years the land surface of Africa has been worn to an almost flat pediplain then, following isostatic uplift, a new cycle of parallel scarp retreat and pediplain extension began to extend inland from the coast to create a new planed surface (Fig 61). The remains of each pediplain thus exist at different elevations, the highest surfaces being the oldest pediplain.

In Africa the oldest and highest surface is called the Gondwana surface, pediplanation having probably occurred when Africa was part of the Gondwana landmass during the Jurassic and early Cretaceous. When Gondwanaland broke up, erosion had reduced the surface to about 600 m above sea-level. Then extensive coastal regions sank slowly beneath the seas and vast thicknesses of sediment were laid down. The post-Gondwana cycle of erosion began to wear the interior continental surface down further, delivering vast amounts of sediment to the continental shelf. The weight of these offshore sediments caused isostatic uplift of the continent, warping its surface and tilting the Gondwana and post-Gondwana surfaces. The third or African cycle now began and continued throughout the Tertiary period until the early Miocene period when further isostatic uplift occurred. The African cycle was followed by the post-African cycle and together these form the most extensive erosion levels or bevels in Africa. During the Pliocene a fourth cycle, the Victoria Falls cycle, began and was followed by the Congo cycle in the Pleistocene period. Since the end of the Pleistocene another cycle of erosion appears to have made minor inroads along the southern edges of the continent.

Southern Africa clearly has several landscape levels representing different erosion cycles, although tilting, folding and warping during uplift have complicated the pattern. Because of the greater uplift of the southern and eastern parts of the continent, the erosion surfaces decline in altitude northwards and westwards and the vertical interval between each planation bevel is also reduced. In the northern parts of West Africa the planation surfaces are impossible to trace for they have been buried under vast thicknesses of Cretaceous and Tertiary deposits.

In *South Africa*, the Great Escarpment divides the older erosion surfaces of the interior from the younger surfaces of the coastal regions which are extending inland. It is therefore an excellent example of an erosion scarp, an elevated rim of Gondwanaland. The Great Escarpment runs from Victoria Falls across Zimbabwe to Marandellas then southwards along the easter edge of the Zimbabwe highveld to the Limpopo. It re-appears in the Drakensberg of the northeastern Transvaal where the Letaba, Olifants and other rivers are gradually eating

Fig. 59. Zimbabwe: looking north-west over the Lomagundi Plateau—an erosion plain of the African cycle—from the Umvukwe Range, the northern section of the Great Dyke, north-west of Salisbury

Fig 60. Nigeria: this photograph, taken from the top of a tor, shows the High Plains of Hausaland near Zaria, which have been cut across granites and gneisses. It is an erosion pediplain of L.C. King's African or early Tertiary cycle

Fig. 61. Block diagrams illustrating the chief principles of the theory of pediplanation: (a) First phase of erosion: rivers and streams eat laterally into the landscape causing the retreat of scarps and the extension of pediplains. Waste material is transferred to the continental shelf. (b) Second phase of erosion: the continent is up-lifted due to isostatic readjust-ment. While the first erosion phase continues at a higher elevation, a second erosion phase begins to extend pedi-plains at a new lower level inland from the coasts

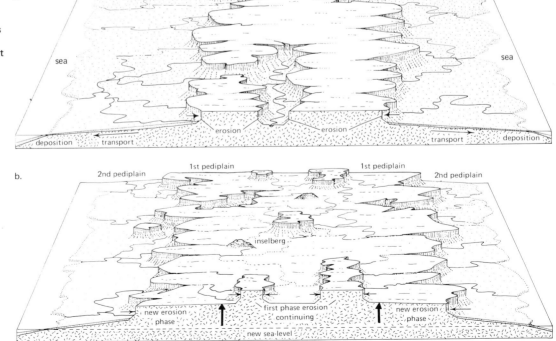

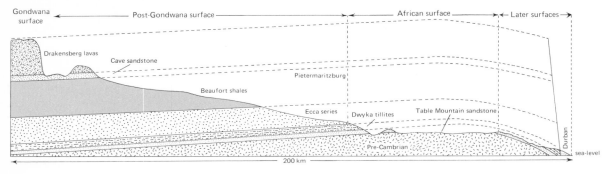

Fig. 62.   Section through Natal, South Africa from the Drakensberg to Durban. The Great Escarpment, left, has retreated nearly 200 km since Jurassic times

into the post-African surface north of Pretoria. In the Lydenburg district volcanic rocks add to the height of the quartzite scarp (Mt Anderson, 2285 m) and here the Incomati and Blyde are eroding the scarp backwards.

The Natal-Lesotho region, 365 km long, is the most impressive portion of the Great Escarpment. Here resistant Stormberg lavas cap the softer Karoo sandstones and shales, the whole rampart rising to over 1400 m and often to over 2000 m (Fig 62). This is the highest part of southern Africa with many prominent peaks—Mont aux Sources (3299 m), Champagne Castle (3376 m) and Giant's Castle (3313 m). The Escarpment then curves in a 675 km-long arc in the Stormberge, Suur and Sneeuwberge, the Nuweveldberge and the Roggeveldberge. Here the Brak, Kariega, Salt and Sundays are eroding the face of the Escarpment.

Fig 63.   South Africa, the Drakensberg Mountains, Natal: over 1 000m of horizontally laid basaltic lavas overlie Table Mountain Sandstone along this erosion scarp. The foothills show evidence of surface creep

In its southwestern section the Escarpment becomes less distinct, continuing in the Langeberg and in the granite highlands of Namaqualand. North of the Orange, it reappears in the Huns Mountains, the Schwarzrand, the Zaria Mountains and the Khomas Highlands where it separates the Namib Desert from the plateau interior.

Inland from this extensive escarpment lie the interior plateaux. Nearly half this area is composed of rocks of Karoo age which have been downwarped into a huge basin. The landscape here is one of extensive pediplains broken by steep-sided koppies capped with resistant bands of sandstone and dolerite (Fig 66). Several erosion surfaces can be recognised on the interior plateaux: The Gondwana surface and the post-Gondwana surface form the summit plateau of the Lesotho Highlands between 2575 m and 3200 m and the tops of the Chimanimani Mountains of eastern Zimbabwe. African surfaces appear as smooth level plains at about 1800 m on the Kaap Plateau, the interior plateau of Namibia and on the highveld of the Transvaal between Pretoria and Middleburg. Post-African surfaces are very extensive and dominate the Namib, the Transvaal Bushveld, much of Botswana, the Cape Middle Veld, northern Transvaal and occur in the Mrewa, Matoko and Matopas regions in Zimbabwe and where the Shashi and Limpopo rivers meet. These surfaces extend into Zambia and Angola.

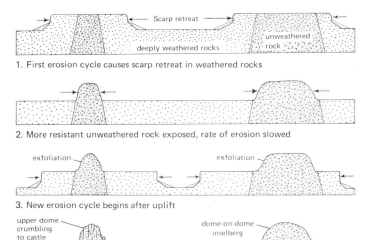

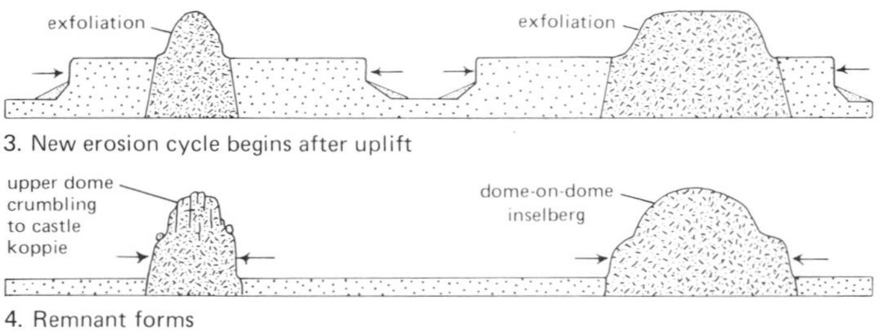

1. First erosion cycle causes scarp retreat in weathered rocks

2. More resistant unweathered rock exposed, rate of erosion slowed

3. New erosion cycle begins after uplift

4. Remnant forms

Fig 64.   Landforms in resistant rock due to uplift causing a second erosion cycle

Fig. 65. South Africa, near Steynsburg, northeastern Cape Province: typical inselbergs or koppies of the Karoo at different stages of erosion. On the right, a flat-topped koppie called the Koffiebus or Coffee Jar. This will eventually be eroded to a spitzkoppie like that on the left called the Teebus or Tea Jar. The final stages are seen in the low hills. Layers of basaltic lava are protecting the underlying sandstone

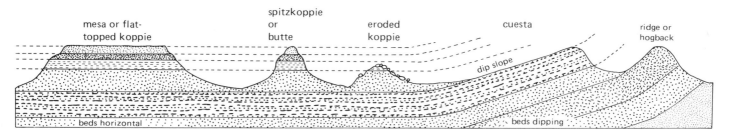

Fig.66. Some landforms due to erosional processes

Fig. 67. Malawi: Zomba Mountain in Southern Province is a coarse-grained igneous intrusion which rises to a plateau at 1 500m to 2 100m. King considers this plateau surface to be a remnant of his Gondwana erosion level. In the foreground is part of the extensive pediplaned surface of Central Africa at 1 000m. This is L.C. King's African planation surface.

43

Fig. 68. Uganda, Bunyoro Province: an erosion level at about 1 000m in the lower Victoria Nile area. The Nile is seen (left) just after it has left the Kabalega Falls

King has linked the southern Africa erosion surfaces with those of East Africa. In Kenya Gondwana and post-Gondwana bevels occur at approximately 3000 m on parts of the Aberdares and the Mathew Range, while the African cycle is seen on the Turkana Plains and the Masai Plains west of the Rift Valley, between the Uaso Nyro and Mara rivers at about 1700 m, but here it is masked in parts by lava flows and interrupted by faulting. The surface extends into the higher Serengeti Plain of Tanzania.

In Uganda, Gondwana erosion surfaces (here called the Ankole surface) exist in the western Rift Valley highlands between 1820 m and 1970 m and form rolling surfaces (Fig 69). The post-Gondwana level (Koki surface) stands between 1120 m and 1450 m in the Masaka region. The African surface is seen on small hills above the post-African plains at about 1515 m in Ankole but towards the west this surface has been greatly dissected and in Buganda forms lateritic flat-topped hills at about 1330 m separated by swampy valley floors at about 1120 m. A later surface, the Acholi surface, was eroded during the Pleistocene and stands between 910 m and 1000 m.

In *West Africa* the Gondwana surface is seen on the summits of the Idanre Mountains of southwest Nigeria, and in the Adamawa Highlands (Fig 70), on Mt Agou in Togo, and on the Nimba Mountains of southeastern Guinea. The most widespread pediplaned surface in West Africa lies between 425 m and 610 m and is separated by steep scarps from a second surface at 730 m. Most of West Africa's planation surfaces have been gently warped making recognition of the full sequence from the Gondwana to the present cycle difficult. Six major levels of planation are recognised, although the closeness of the bevelling causes a confused combination of three or more cycles.

Fieldwork

Measurement of Slope: A simple clinometer for measuring the angle of slopes can be made by attaching a piece of thin stiff plastic tubing or a strong drinking straw to the straight edge of a plastic or wooden protractor. Bore a small hole in the centre of the base of the protractor and suspend through it a length of cotton with a small weight attached to the end. The observer sights through the tube up-slope to the top of a ranging pole the same height as the observer. The weighted cotton will indicate an angle on the protractor which, when subtraced from 90°, will give the angle of slope.

1. What is the importance of slope study in geomorphology?

Map Work

Slope Analysis: Slopes on contour maps can be analysed by noting the density of the contour lines and classifying these densities according to slope type—slight, moderate, steep, etc. Take a strip of straight-edged paper, align it along the linear scale of the map, and mark off distances of 100 m, 500 m and 1000 m. Calibrate the straight edge so that 100 m equals a slope of 1 in 2, the 500 m equals 1 in 10, and the 1000 m equals 1 in 20 (if the contour interval is 50 m). By placing this scale between the contour lines you can calculate the linear distance taken to rise 50 m. If the linear distance is 400 m between adjacent contour lines of 50 m interval, then the gradient is 50 m rise in 400 m linear distance or 1 in 8. Large areas of contour maps can be graded quickly by degree of slope into steep slope, moderate slope, etc., and shaded accordingly. Very flat regions will stand out in one colour of shading and will indicate possible

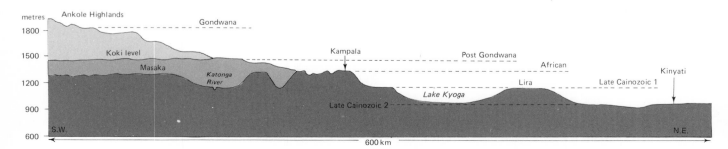

Fig 69. Three diagrammatic cross-sections showing the various erosion levels in Uganda

Fig 70. The distribution of the Gondwana erosion surface in Nigeria and Cameroon.

Questions

1. What does the term 'base-level' mean? Discuss the importance of the base-level concept in the study of landscape development.

2. Outline the main features of W. M. Davis's geographical or landscape cycle. Why is this theory not a suitable one for the explanation of landscape development in Africa?

3. Outline W. Penck's pediplanation theory as adapted by L. C. King to explain the development of landscapes in Africa. What evidence is there to support this theory?

4. What is climatic geomorphology? What is dynamic equilibrium? How do these concepts help us to understand the development of landscapes?

5. Write an essay entitled 'The Erosion Surfaces of Africa'. You should refer to any evidence of erosion surfaces in your own country and to map work or personal survey you have carried out which indicates the presence of erosion levels in your region.

6. Outline the process of planation by deep weathering as described by J. Büdel and the planation process as described by L. C. King. In both cases you should give examples of planed surfaces in Africa.

planation levels.

An alternative method of showing degree of slope is to super-impose a network of squares on to a map (say 2 km squares on a 1:50 000 scale). Grade the squares by numbering them 1, 2, 3, etc., according to the number of contour lines falling within the squares, e.g. a square with ten lines crossing it may be graded 5, one with 8 as a 4 and so on. Those with no contour lines or, say only one or two, will be graded 0 and these squares may indicate possible planation levels.

Fig. 71. Kenya, Tsavo West National Game Park: plains cut across metamorphic rocks. They are late Tertiary in age and accord with King's post-African cycle

8 Riverine Landscapes

A stream or river is a body of water which transports dissolved substances and particles or rock and which flows down a slope and along a well-defined channel. The study of the work of a river involves the following aspects:

(a) *discharge*: the amount of water which flows past a certain point during a measurement of time. This can be measured in cubic metres per second ($m^3 s^{-1}$) or in cubic feet per second.

(b) *energy*: the power of the stream to move its load

(c) *channel development*

(d) *load*: its transport and deposition

(e) *river profile or gradient*: the slope measured along the bed or surface of the river from source to mouth

(f) *effects of base-level change*: by the relative change of sea-level relative to land. This may be due to land movement (isostatic change) or sea-level change (eustatic change).

(f) *adjustment to structure*.

Streams and rivers are the most important element in landscape development. While only limited regions of the earth's surface have been moulded by glaciation and wind, vast regions consist of valleys and hills or interfluves—the work of streams and mass wasting. Streams are an essential part of the water cycle, moving millions of tonnes of water from land to sea each year. They are important geological agents transporting each year about 400 million tonnes of material in solution and 1000 million tonnes mechanically from the earth's surface to the oceans.

Rain falling on the ground which is not returned to the atmosphere by evaporation or transpiration percolates or seeps into the soil and rock or flows over the surface as run-off. The run-off may flow across the land surface in broad sheets as surface wash or sheet flow, or it may flow in a well-defined channel as a stream. Sheet flow occurs where the ground is unable to absorb all the rain falling on it and is responsible for sheet erosion.

Discharge of Africa's Major Rivers

Measurements taken at gauging stations along Africa's major rivers show wide seasonal fluctuations in discharge rates reflecting the effects of alternating dry and rainy seasons. Discharge of a river can be expressed as

$$Q = w \times d \times v$$
$$\text{Discharge} = \text{width} \times \text{depth} \times \text{velocity}$$
$$(m^3 s^{-1})$$

When discharge increases, erosion and channel enlargement increase, when discharge decreases some of the load is dropped reducing the channel depth and width. The channel of the stream or river is thus linked closely to the discharge; if the discharge changes, width, depth and velocity must change also.

Figs 73 and 75 show the extreme rates of discharge of some of Africa's rivers. Other examples include the Volta in Ghana which has a dry season discharge of $40\,000\,m^3 s^{-1}$ compared

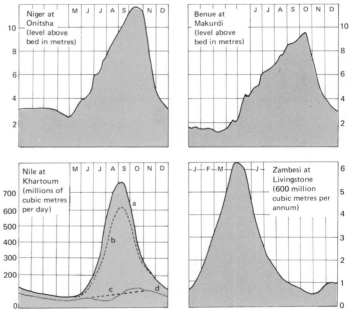

Fig 73. Seasonal variation of water levels and discharge on four major rivers. In the Nile diagram: (a) main Nile below R. Atbara; (b) main Nile at Khartoum; (c) contribution from Blue Nile; (d) contribution from White Nile.

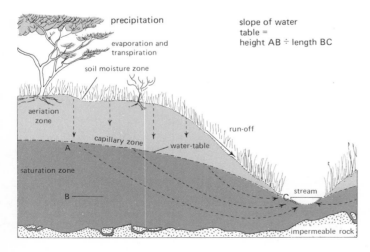

Fig 72. Water movement and distribution in the soil and regolith. The presence of lateritic layers may alter this pattern.

Fig 74.   South Africa: a tributary of the Oliphants, south-west Cape Province. The river shows its deposited load, slip-off slopes and under-cutting of banks. Photograph taken during the dry summer

with a wet season $122\,000\ \mathrm{m^3 s^{-1}}$ and its level fluctuates by as much as 12·5 m in a year. The Orange at Prieska sometimes has almost no flow in October at the end of the dry season compared with a maximum of 6·6 million $\mathrm{m^3}$ flowing in March. The flow of the Zambesi at Livingstone drops from a maximum of 11·25 million $\mathrm{m^3}$ in March to a minimum of 0·42 million $\mathrm{m^3}$ in October. Of Africa's major rivers, only the Zaïre maintains a constant high volume because of its huge catchment area. At the Inga Falls 80 km upstream from Matadi the Zaïre has a flow of $22\,700\ \mathrm{m^3 s^{-1}}$ at low level and $60\,900\ \mathrm{m^3 s^{-1}}$ at flood—a flow second only to the Amazon—and its fresh waters form a surface current reaching 80 km out into the Atlantic.

Fig 76.   Tanzania: a stream flowing over a volcanic dyke, lower slopes of Kilimanjaro. Notice the variety of pebbles and stones in the river bed. Are the largest boulders part of the river's load?

Fig 75.   Fluctuations on a smaller African river: the Tana River Kenya. The two most variable years, 1953 (pecked line) and 1965 (solid line) are compared.

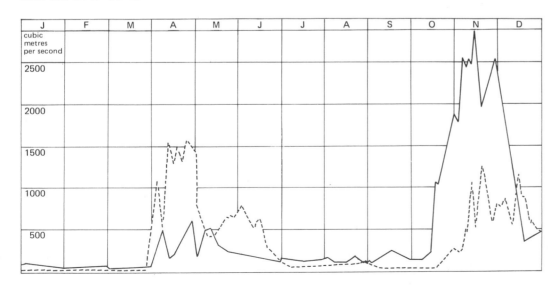

Stream Energy and Capacity

This marked seasonal discharge of Africa's rivers means that much of their erosive and carrying capacity is concentrated in only part of the year. A stream's energy is used to carry its load but a considerable amount of energy is lost because of friction between the flowing water and the bed and sides of the stream channel. The smaller the volume of water in the channel, the more energy is lost in proportion to friction. With higher velocities, the stream's capacity to move a given load increases and a doubling of speed can cause a sixty-fold increase in stream capacity. The width and depth of a river's channel, however, has a great influence on a river's capacity (Fig 77).

If one examines the visible load of a stream, it is seen to be made up of pebbles and stones of various sizes and the diameter of the largest stone or particle gives us a measure of the ability of the river to transport its load. This is termed the competence of the river (Fig 78). The load can be carried in three different ways. Chemically weathered salts such as chlorides and carbonates are moved in solution. These salts are largely contributed by ground-water seepage and are not visible (p. 21). Light rock particles such as silt and clay are carried in suspension and give the stream its muddy appearance. Larger pebbles, stones, cobbles and small boulders—the bed load—are rolled, bounced or jostled along the bed, a process called traction. With increased velocity particles usually moved by traction may be lifted in suspension. As the stones move they become worn or abraded and lighter and smoother as the river progresses downstream until in the lower reaches the load may be entirely composed of fine particles in suspension and salts in solution. When the river's velocity falls, the river's competence decreases and the heavier particles are dropped. They will not be picked up again until the river reaches the lowest speed required to move particles of that weight and size—the critical erosion speed. Surprisingly, this speed is lower for sand particles than for smaller clay particles. This is because the clay particles present a smoother surface to the water flow and are not easily picked up.

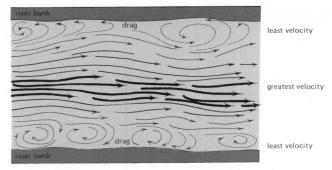

Fig 78. Velocity variations in a river channel due to drag

Channel Development

The movement of the traction load along the river channel deepens and widens it by friction. This process, termed corrasion is the main way in which the valley is widened and deepened: the greater the speed and the heavier the load, the wider and deeper the river channel. The valley sides are worn back, however, by surface wash and wasting (Fig 80). Chemicals carried in the water may react with rock chemicals as in limestone and help in channel development, a process called corrosion.

The water alone can also help channel enlargement by hydraulic action—the loosening and removal of rock particles by water pressure. This pressure is not evenly applied due to turbulence in the water flow caused by irregularities on the bed and sides, higher speeds in the centre of the water flow where friction is absent, and by lower speeds in deeper sections (Fig 78).

Occasionally stones are caught in small depressions on the river bed and are swirled round to erode deep hollows called

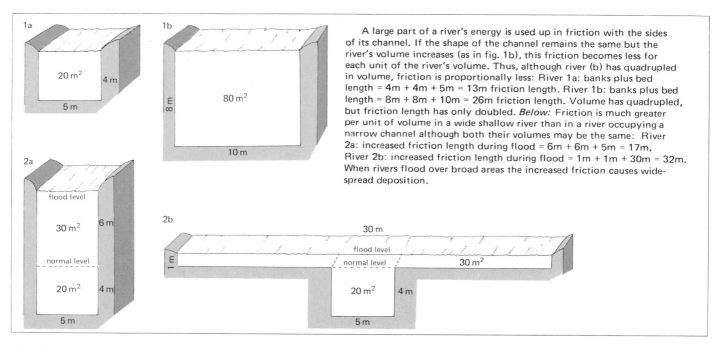

A large part of a river's energy is used up in friction with the sides of its channel. If the shape of the channel remains the same but the river's volume increases (as in fig. 1b), this friction becomes less for each unit of the river's volume. Thus, although river (b) has quadrupled in volume, friction is proportionally less: River 1a: banks plus bed length = 4m + 4m + 5m = 13m friction length. River 1b: banks plus bed length = 8m + 8m + 10m = 26m friction length. Volume has quadrupled, but friction length has only doubled. *Below:* Friction is much greater per unit of volume in a wide shallow river than in a river occupying a narrow channel although both their volumes may be the same: River 2a: increased friction length during flood = 6m + 6m + 5m = 17m. River 2b: increased friction length during flood = 1m + 1m + 30m = 32m. When rivers flood over broad areas the increased friction causes widespread deposition.

Fig. 77. Effects of change in channel depth and width: (a) Friction between water and channel sides decreases as river increases in size if channel shape remains constant. (b) Effects of equal volumes of flood water in a narrow channel and a broad channel

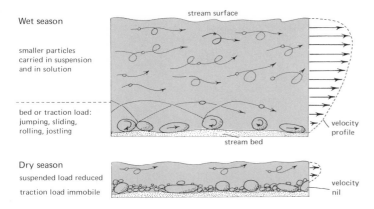

smaller particles
carried in suspension
and in solution

bed or traction load:
jumping, sliding,
rolling, jostling

stream surface

velocity
profile

stream bed

Dry season
suspended load reduced

traction load immobile

velocity
nil

Fig 79.   Transport of a river's load during wet and dry seasons

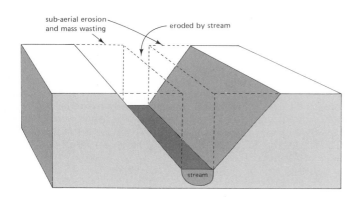

sub-aerial erosion
and mass wasting

eroded by stream

stream

Fig 80.   Block diagram showing the much greater amount of rock eroded by mass wasting and sheet erosion than by a stream. The stream, however, is responsible for transporting the total debris

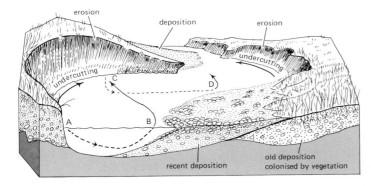

erosion

deposition

erosion

undercutting

undercutting

C

D

A   B

recent deposition

old deposition
colonised by vegetation

Fig 81.   Lateral extension of river meanders by deposition and undercutting. The corkscrew flow in a meander is indicated by the movement from A to D.

Fig 82.   South Africa, Burke's Luck, Blyde River Canyon: potholes eroded in dolomite. The railings (top right) give the scale

pot-holes. This action lowers the bed rapidly, especially in soft rocks, and often forms intricate patterns (Fig 81).

Continuous corrasion in arid regions where vertical erosion exceeds lateral erosion will eventually produce a gorge or canyon; the Fish River Canyon (Fig 83) in Namibia is a good example as are many of the deeper wadis (p. 112) in the African deserts. Normally, however, lateral corrasion and sub-aerial weathering of the valley sides widens the valley. Lateral corrasion is particularly active where the current is swept to the outside of bends (Fig 82) to undercut the river bank. The inner bank or slip-off slope where the water is slower is an area of minimum erosion and of deposition (Fig 84). By undercutting and depositing at bends the river develops ever wider meanders. Eventually the 'neck' of the meander may be cut through and an ox-bow lake is formed. Ox-bow lakes are ephemeral features, soon drying up into curves of marshy ground or a belt of more luxuriant vegetation.

The size of meanders, that is, their wave-length and amplitude (Fig 84) is proportional to the volume of water in a full river channel and the radius of the curvature is two or three times the width of the stream channel. Meanders are not formed accidentally but are most common in fine-grained alluvium. They are curves in the river's course which are related to the accomplishment of the minimum amount of effort. The width of the belt occupied by the meander channels is gradually extended into the valley sides and is usually about twenty times the width of the river's channel.

Deposition of Load

Deposition occurs when the river's speed falls below the minimum required to transport the particles. This can be caused by the river entering a lake or sea, by high evaporation rates causing water loss, by a decrease in gradient, or by a widening of the stream bed causing increased friction.

A sudden check in speed occurs where a stream leaves a highland mass to flow onto a lowland plain. Here a large half-cone fan of debris, an alluvial fan, will build up; alluvial fans are a common feature in arid and semi-arid regions (see p. 112). Deposition also occurs periodically when a river overflows into the flood plain. Vast sediments are built up due to the reduction in speed in the shallowing water, e.g. the banto faros along the river Gambia, the fadamas of Hausaland, the walo of the Senegal, the hofrats of the Mauritanian Sahel zone and the flood plains of the inland Niger delta. Low ridges called levées (Fig 85) are built up along the banks of the river where the heavier debris is deposited, for example, along the Nile, and the lower Omo River where it enters L. Turkana. Deposition also occurs on the river bed itself, especially towards the end of the rainy season when volume and speed are falling, and in areas of lessening gradient. The river then breaks up into numerous braided channels flowing between low sand banks, for example, on the wider stretches of the Limpopo and Zambesi and the lower Senegal. The Niger in its inland delta breaks up into several channels as in the Marigot de Diaka district where the gradient falls to 1:17 000. Similar braided channels occur on the Orange River just above the Aughrabies Falls to form large islands—Klas, Tierhof and Ox islands (Fig 102). The Kilombero and Rufiji in Tanzania, and the Omo in Ethiopia have braided channels and the Mkomanzi in Tanzania displays a miniature inland delta similar to that of the Niger (map 11).

When the river enters a sea, lake or swamp, its speed falls and deposition is immediate. If there are shallow waters, no strong currents, and a large and continuous supply of sediment,

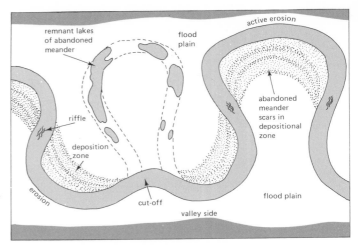

Fig 84. Some major features of a river's meandering channel. Riffles are more turbulent water where the river current crosses over from one bank to another between bends (see Fig 81)

a delta may form. The quiet, still waters of lakes are ideal for delta development: the River Omo has built a bird's foot delta some 30 km wide at the northern end of L. Rudolf (Fig 93) and the Semliki has extended a delta into the southern end of L. Mobutu. A lake delta in the process of formation can be seen at the northern end of L. Malombe in Malaŵi where the Shiré River deposits silt carried from the southern end of L. Malaŵi. In western Kenya the Kano Plains at the eastern end of the downfaulted Kavirondo Gulf are being gradually extended by sediments deposited by the Sondu and Nyando rivers. Strandlines now well above the present water levels in lakes Elementeita, Naivasha and Natron in East Africa have also revealed evidence of former deltas.

Fig 83. Namibia: incised meanders on the Fish River. In this arid region the river has cut its gorge 450m to 600m deep in the sandstones, shales and limestones of the Nama beds

Fig. 85. Kenya, West Nyanza District: the meandering river Ngaila displays rapids, undercut banks, slip-off slopes, sandbars and double channels

Deltas will continue to grow as long as the supply of sediments is maintained. The sediments are deposited first as bottomset beds of off-shore clays, then as a steeper middle section called the foreset beds of silty clays; and finally as a flat upper portion called the topset beds of silts and sands (Fig 87). The delta becomes fan-like in plan with radiating distributary channels meandering across its surface. The Niger and the Benue, because of the long distances and generally soft rocks over which they flow, have built up a huge delta. This arcuate or cuspate delta stretches 480 km from Opobo to near the Benin River mouth and the sediments are estimated to be 12 000 m deep. The Niger empties into the Bight of Benin at the Nun entrance but numerous other distributary rivers wander over the fan (Fig 89).

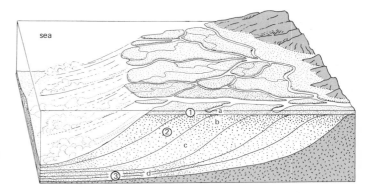

Fig 87. Block diagram of a delta showing: 1. topset beds; 2. foreset beds; 3. bottomset beds. The deposits consist of: (a) marsh silts; (b) delta front silts and sands; (c) silty clays; (d) clays

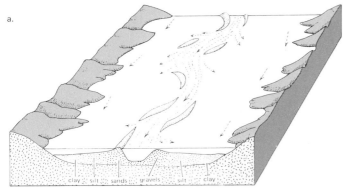

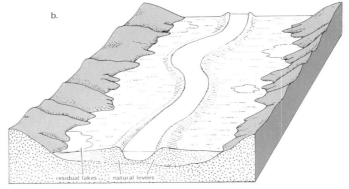

Fig. 86. The formation of levees. (a) River in flood. Water in central channel continues to flow fairly quickly but water spreading into surrounding areas decreases in speed. Coarser silt is dropped near central channel, finer clays further away. (b) During normal levels the river's water is contained between the high banks or levees

51

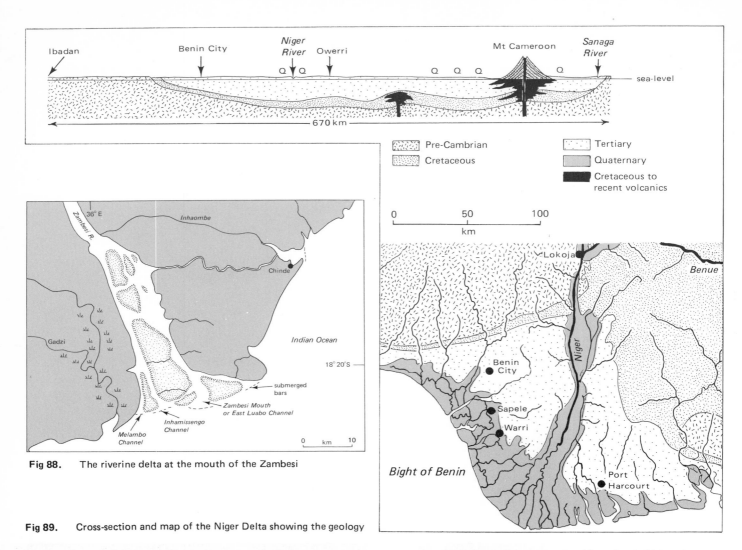

Fig 88.   The riverine delta at the mouth of the Zambesi

Fig 89.   Cross-section and map of the Niger Delta showing the geology

Fig. 90.   South Africa: the Orange River where it leaves the Lesotho Highlands for the Highveld of the Orange Free State. Note the undercutting (left centre), the deposition (centre) and the sandbanks emerging above the shallow river. The photograph was taken when the river's level was low. Where, approximately, is the high level mark?

Fig. 91. South Africa: the Vaal River near Vereeniging is typical of many of Africa's major rivers with its broad shallow channel, low banks and sand-banks exposed at low level

The Nile Delta is a classic example of the triangular arcuate form. Its greatest extent from its apex is 185 km and its coastal arc is 240 km long. The delta surface is crossed by numerous levée-flanked distributaries. The silt from the two major channels, the Rosetta and Damietta courses, has extended tongues of sediment into the Mediterranean. Long sand-spits, curved by sea currents, enclose large lagoons such as the Bahra el Burullus which is 55 km long by 14 km wide.

The Volta in Ghana has also built a flat arcuate delta some 115 km wide between Denu in the east and Akplabanya in the west. Here again, sandspits and bars have blocked many of the distributary mouths and enclosed water bodies, for example,

the Keta and Songaw lagoons. The smallness of this delta in comparison with the size of the Volta River is due to sediments filling in a former ria before appearing at the surface.

In Tanzania, the *Rufiji River* draws its sediments from a catchment area covering 177 500 km^2 and has produced an extensive delta at its mouth. Further south, deltas are not a prominent feature of the coasts due to severe scouring currents. South of 10°S the *Zambesi* is the only river to have a significant delta of the estuarine type and even this is small for such a large river, being only 60 km across. The distributary channels are shallow and often blocked by sand-bars and only the Chinde channel is navigable (Fig 90).

Fig. 92. Egypt: the Nile Delta from the spacecraft Gemini 4 at a height of about 240 km. The two major distributaries have built out levees into the Mediterranean at the Masabb Rashid (mouth of the Rosetta) (bottom) and at Masabb Dumyat (mouth of the Damietta) Between them lies the almost enclosed Lake Burullus

The River Profile

A river may begin as a rushing torrent and end as a slowly meandering flow of turbid water. Normally, three stages can be recognised in a river's course:

(a) *Youthful: torrent or upper stage*: Characterised by a steep gradient, swift-flowing current, vertical erosion, a deep V-shaped cross section, and interlocking spurs. The load consists of large angular particles. This stage of the river's course is seen in the elevated regions of Africa—the Aberdares, the Adamawa and Cameroon highlands, the Sula Mountains of Sierra Leone, the Chimanimani and Inyanga Mountains of Rhodesia, and the western Futa Djallon.

(b) *Mature: valley or middle stage*: Lateral erosion is now more important than vertical erosion; the gradient is more gentle, and a flood plain is emerging; meanders are developing and the load consists of rounded and smaller stones. The upper Niger above the inland Delta and on the Sokoto plains, the upper Benue valley, the Black Volta and the Rufiji's middle course are examples.

(c) *Old: plain or lower stage*: Deposition now exceeds erosion, the river has a very gentle gradient, and the river flows in a wide, flat plain which displays many depositional features. The load is largely silt and salts in suspension.

Most rivers thus display some concavity in their long profiles grading from the steep youthful stage to an almost flat plain in the lower stage. Clearly the river is attempting to erode its long profile down to base-level (p. 37) which may be sea-level or a local base-level such as a pan or lake surface or an outcrop of hard rock. In the case of a local base-level there is an immediate steepening of gradient immediately downstream of the interrupting feature (Fig 93).

None of the rivers of Africa has achieved the ideal graded long profile of smooth concavity. In fact, the long profile of many African rivers would appear to be partly graded concave sections frequently interrupted by lakes, resistant rocks, and volcanic or tectonic features. The sections (Figs 94 and 95) show that the Niger's long profile resembles the ideal.

Fig 93. Resistant rock outcrops and lake surfaces forming local base levels slow down the river's progress in attaining the ideal concave long profile

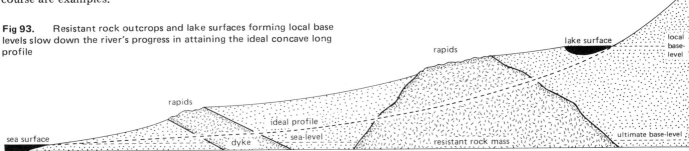

Fig 94. Long profile along the Niger River

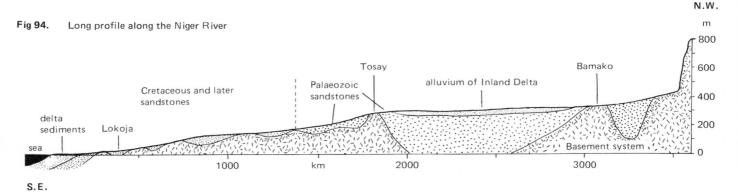

Fig 95. Long profile along the Zaire or Congo River

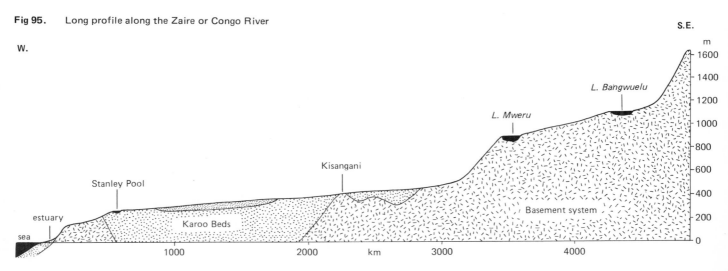

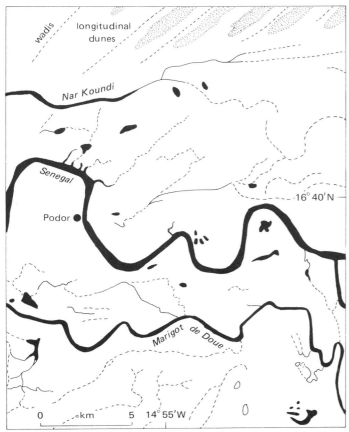

Fig 96. The broad flood plain of the lower Senegal displays double channels, remnant ox-bow lakes and the broad meanders of an old river

Fig 98. Ethiopia: the Omo River has formed several bird's foot deltas where it enters Lake Rudolf. These are the distributaries of the Dielerhiele delta. Note the clearly defined levees (upper left)

Fig 97. South Africa, the Valley of a Thousand Hills east of Pietermaritzburg, Natal: the valleys have deeply incised themselves into the Table Mountain Sandstone and underlying Old Granite. Note the interlocking spurs in the foreground

Waterfalls and Rapids

The interruption of the river's long profile by a resistant band of rock is frequently marked by waterfalls or rapids. The river is unable to erode the outcrops as easily as the adjacent softer beds and the area immediately downstream from the interruption is worn to a lower level to cause a waterfall. Here, the softer rock is scoured deeply by water falling under gravity and a plunge pool forms. Eventually the scouring removes softer rock from beneath the hard rock ledge which, unsupported, collapses into the plunge pool. By this process the edge of the fall migrates upstream sometimes creating a narrow gorge. As the falls retreat they will become lower and develop into rapids until finally the hard band is worn down to the general surface of the river's long profile (Fig 101).

Waterfalls of this type occur in many parts of Africa because of the existence of numerous lava bands and outcrops of hard basement rocks. An almost perfect example is the Howick Falls some 21 km north of Pietermaritzburg in Natal where the Umgeni River plunges over a 60 m-thick dolerite sill to fall 92 m into a narrow gorge (Fig 100). The Maletsunyane Falls 130 km southeast of Maseru in Lesotho are one of the highest in southern Africa, falling 477 m over a ledge of Drakensberg basalt. The Orange River, just above its junction with the Malopo, first crosses massive granite and then encounters foliated beds. This junction is marked by the Aughrabies Falls. Here, after a series of rapids, the Orange drops 155 m into a plunge pool some 43 m deep before continuing its way across the foliated beds in which it has worn a zig-zag course about 16 km long. The river

Fig 101. South Africa, the Howick Falls, Natal: the water tumbles over a dolomite sill overlying sedimentary beds. Notice the deep plunge pool at the base and the stream load (bottom left)

Fig 99. Stages in the recession and removal of a waterfall. The falls have retreated from (a) to (b) by undercutting and collapse of the resistant rock bed. Eventually rapids will be formed at (c) and the river will attain a graded profile at (d). 1 — 4 are the various profiles of the stream at each stage.

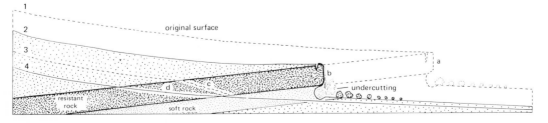

which was 3 km wide above the falls, is now constricted to the gorge 150 m wide (Fig 102). In Zambia the Kalombo River which flows into L. Tanganyika has eroded a deep gorge and created the Kalombo Falls over 200 m high (Fig 104). Two of the highest waterfalls in Africa, the Mutaruru and Mtarazi Falls, plunge their way over the vertical basalt-capped edge of the Inyanga escarpment in Zimbabwe. The Thika Falls in Kenya are also due to the Thika River flowing over a volcanic dyke (p. 96) and there are many other magnificent falls of similar type in the Kenya Highlands, e.g. Thomson's Falls and Broderick Falls.

The Niger is interrupted by many falls and rapids; in the upper basin where it crosses the Kouroussa dyke, below Bamako where the Niger flows over sandstone outcrops of the Manding Mountains, and over the basement outcrops at Fafa, Labbezenga and Bussa. The Benue also has several rapids in its upper course. The granite escarpment of the Jos Plateau also gives rise to numerous falls, the most notable being the Kurra Falls on the Kurra River. Navigation of Ghana's rivers is also hampered by falls caused by rock outcrops, for example, the Bosomasi rapids on the Pra, and the Senchi rapids of the Volta 8 km north of Kpong.

Fig 102. South Africa, the Aughrabies Falls on the Orange River: note the braided channels above the deep narrow gorge. This is a nick point of the Coastal erosion surface eating inland into the African erosion surface

Fig 103. Uganda, the Kabalega Falls on the Victoria Nile: the falls were formed by the overflow of water from Lake Victoria via the Lake Kyoga drainage system

Fig 104. Tanzania-Zambia border: the Kalombo Falls lie near the south-eastern shore of Lake Tanganyika. They fall 214m and are the highest falls of a series which have cut a 5 km-long gorge from the lake (773m) to the plateau surface, here at about 1 500m

The Effects of Base-level Change

A lowering of the base-level by either continental uplift (isostatic change) or by a drop in sea-level (eustatic change) has an immediate effect on drainage systems. The increase in head of water from the river's source to its mouth increases the gradient of the long profile which, in turn, increases the speed, erosive powers and carrying capacity of the river. The rejuvenated river begins to erode a new valley within the old one which migrates slowly up-river from sea-level to create a valley-in-valley form (Fig 104). Thus a young valley may develop within the confines of a broad flood plain of the original old section. A steep change of gradient occurs where the headwaters of the rejuvenated stream are eating backwards along the old long profile;

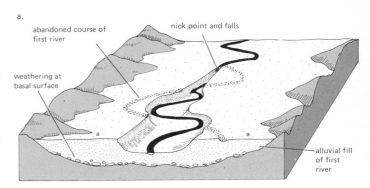

Fig 105. Rejuvenation, valley-in-valley form, and terrace formation: (a) Uplift of land surface causes rejuvenated river to migrate inland along alluvial fill of original river valley. First set of terraces forming at a. (b) Uplift occurs a second time causing second rejuvenation of river and repeat of valley-in-valley form. Two levels of river terraces are seen at a and b

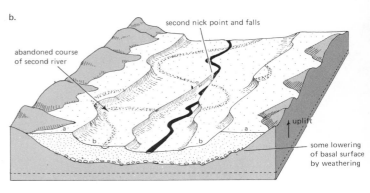

Fig 106. Zambia: (a) view of the Victoria Falls taken from the Zambian side of the Zambesi during the dry season; Zimbabwe in the background. The basaltic cliff over which the river pours is 106m high. Rainbow Falls and Eastern Cataract can be seen in the foreground, Main Falls near centre, and Devil's Cataract partly obscured by spray. The Zambesi exits via the Second Gorge (left)

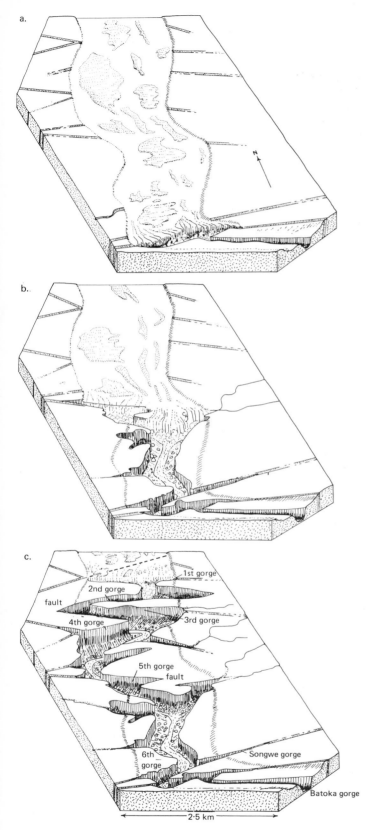

a.

b.

c.

1st gorge

2nd gorge

fault

4th gorge

3rd gorge

5th gorge

fault

6th gorge

Songwe gorge

Batoka gorge

2·5 km

this is the head of rejuvenation or the nick point. The advance up-stream and widening of the new valley is relatively swift because the rejuvenated river is working in the old river's sediments and rotted rocks. As the new valley widens, the old valley surface is reduced to terraces and, if rejuvenation is repeated, a third valley will form in the second to produce a second set of terraces so that the valley sides begin to take on a stepped appearance. Such valley-in-valley forms and terraces are seen in Buganda, on the Nyando River as it crosses the Kano Plains in Nyanza Province in Kenya, and on the Ruzizi near Mbarara in Kigezi Province in Uganda.

In more arid regions where there is little surface water, rejuvenation may cause the deeper incision of the meander pattern of a through-flowing stream to form incised meanders as in the sandstone rocks on the upper Gambia and in the Fish River of Namibia (p. 50). The Fish, some 50 km north of its confluence with the Orange, has cut a deep gorge and covers the last 35 km of its course in a series of deeply incised meanders 5500 m below the general surface.

The uplift of southern Africa during the Tertiary and early Quaternary periods caused rejuvenation along the valleys of the major rivers. Since that time rejuvenated valleys have migrated far into the interior and, where rock structure has been favourable, waterfalls have developed at the nick points. The Cunene River in Angola is a slow-flowing mature to old river in its upper course with numerous meanders and ox-bow lakes and a slight gradient of 1:17 000. At Olushandja some 290 km from the coast the river enters its more youthful valley, rejuvenated by uplift of the plateau rim of Africa. Here it cuts through sandstone via the Caxambue Rapids, the present nick point, its gradient increasing to 1 in 4 and then, 35 km down-stream, plunges over the Ruacana Falls. These falls have been cut back in a deep gorge by the rejuvenated river from the Serra da Cana fault-scarp. Some 150 km from the sea the head of a second youthful valley is marked by the 35 m-high Epupa Falls below which the Cunene cuts through the Baynes Mountains in a rugged gorge. This lower youthful valley is due to an increased water volume caused by river capture (p. 64) rather than to uplift.

Other falls which are due to up-river migration of nick points caused by uplift and initiation of new erosion cycles include the Victoria Falls (Fig 106), the Pungwe Falls on the Zimbabwe Moçambique border (Fig 108), the Aughrabies Falls on the Orange, the Chibirira and Selowandoma gorges on the Sabie and Limpopo rivers in Zimbabwe, and the gorges and rapids lying on the Zaïre between Kinshasa and Matadi.

Fig 107. The development of the Victoria Falls. The Zambesi here flows over horizontally-laid basalt lava which has been weakened by vertical joints and faults (second and fifth gorges). The river took nearly half-a-million years to cut the 18 km of gorges. Height of present Falls: 106m; length: 1860m. (a) Rejuvenated part of Zambesi migrates up-river along faulted Batoka Gorge, encounters first line of weakness, a joint, and erodes first falls. (b) As rejuvenated section moves up-river, subsequent falls are formed along joints and faults, then abandoned. (c) Present position. Seven falls have been abandoned, the present is the eighth, while there are six gorges. The pecked line indicates the future alignment of the ninth falls

Fig 108. Zimbabwe-Mozambique border: the Pungwe River looking upstream from point X on Fig 123. Here the Pungwe is eroding the plateau surface at about 5 500m and then flows into a mature valley. The nick-point falls are in the bottom right-hand corner

Fig 109. Zimbabwe-Mozambique border: the Pungwe Gorge looking downstream from point X on Fig 123. The Pungwe and its tributaries (see escarpment at left) are the agents of an erosion cycle which is gradually extending inland

Fig 110. Terraces on the Zambesi above the Victoria Falls. Vertical scale greatly exaggerated

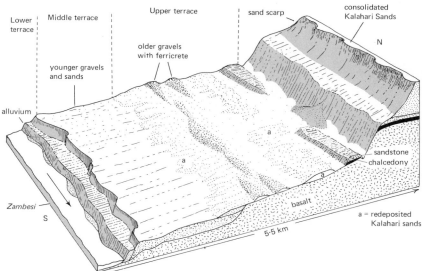

River Adjustment to Tectonic Movements

Where river systems have adjusted to rock structure and surface morphology, their pattern is easy to interpret. Volcanoes such as Kilimanjaro, Mt Kenya and Mt Cameroon all give rise to a typical radial pattern; faulted landscapes possess rectangular patterns as seen in the drainage network on the floor of the Kenya rift valley; tilting of horsts and folding of strata as in Tanzania and the Cape region produce a trellised pattern. Where steep tilting of horsts occurs, e.g., in Kenya and southern Tanzania, many of the rivers are rejuvenated and the degree of tilt can be seen by the alignment of the old terraces with the new valley profile. If tilting is upstream but less than the original gradient, the river will still maintain its flow but at a slower rate and widespread deposition will be caused, e.g. on the middle course of the Mkomanzi River, in the Wembere and Kilombero depressions of Tanzania, along the course of the central Tana in Kenya and in the inland delta of the Niger.

In some cases, however, the amount of back-tilting exceeds the former river gradient and a reversal of the river's former drainage pattern occurs. The classic example of drainage pattern reversal as the river falls to adjust to warping and tilting is seen in the Lake Victoria-Kyoga region of East Africa (Fig 113). The major streams of this region, the Kafu, Katonga and Kagera, originally flowed westwards but upwarping in the west and east and downwarping in the centre caused the Katonga and Kagera to reverse their directions and drain into the depression now occupied by Lake Victoria. The backtilting north of the L. Victoria depression caused the ponding back of the Kafu headwaters. As L. Victoria expanded it overflowed northwards to the Kyoga river channels which were drowned by the excess water and sluggish westerly flow to become a huge inland ria. Headward erosion by a river draining into the northern end of L. Mobutu linked lakes Kyoga and Victoria and the Kafu River with the Nile system. Kabalega Falls were thus formed by overflow waters cutting through a rocky ridge (Fig 103). The Mpanga or

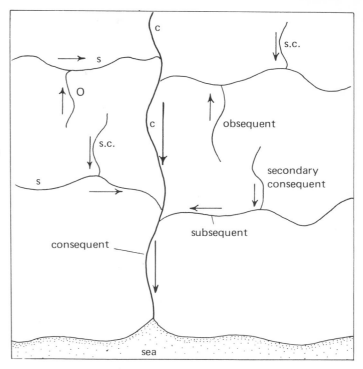

Fig 111. The development of a river pattern

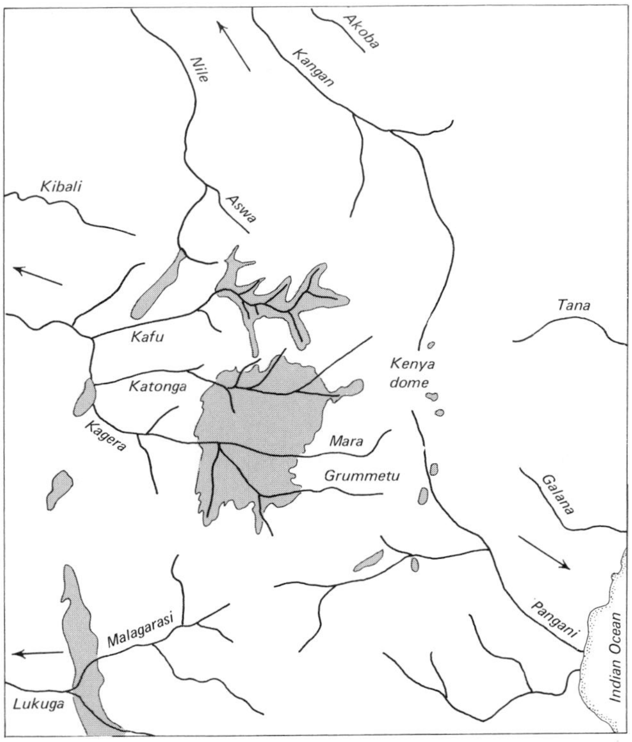

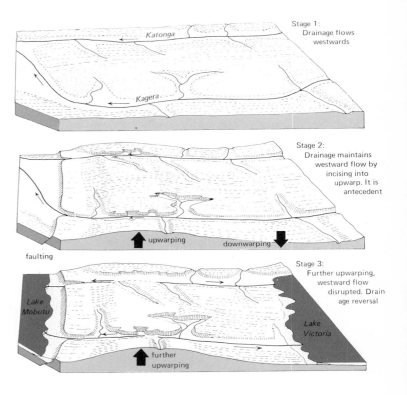

Fig 113. Three block diagrams to illustrate the reversal of drainage and general landscape development due to warping in the western part of Uganda

Fig 112. The drainage of East Africa before the beginning of the Pleistocene period. The position of lakes formed during the Pleistocene is indicated

Rusangwe and the Ntungu or Birira rivers now rise in the same swamps as the eastward-flowing Katonga and Kagera, but they flow westwards to lakes George and Amin as remnants of the original drainage pattern.

At first it will be seen (Fig 113) that the westward-flowing rivers were able to maintain their original or antecedent courses through the rising warp since their rate of downcutting was greater than the rate of uplift; they cut across the structure and became discordant. The Niger has also maintained a similar antecedent course just south of Kainji lake where the river crosses an axis of crystalline rocks running NE–SW between Kontagora and Kaiama in Nigeria by cutting a 150 m-deep gorge. Further upstream, about 120 km southeast of Niamey in the Niger Republic, the Niger course has been maintained once more by a series of zig-zag gorges cut through the end of the Atacora Mountains.

The course of a discordant river is not always antecedent. The Congo/Zaïre between Matadi and Kinshasa, and the Nile at Kabalega Falls in Uganda, overflowed from lakes and cut across local structures. They are therefore not antecedent.

Discordant drainage is also due to superimposition in which a river develops its course on a land surface following its rock structures and weaknesses; as it erodes these upper, younger

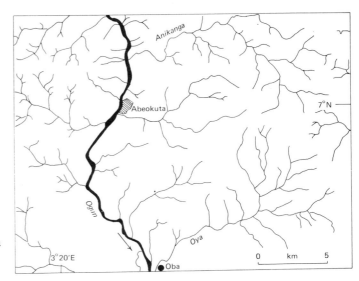

Fig 115.   A dendritic drainage pattern: the Ogun River and its tributaries in the Abeokuta region of southern Nigeria

Fig 114.   Superimposition of the Gouritz-Dwyka-Gamka drainage system, South Africa (for location see Fig 125 and map 10, p. 138.)

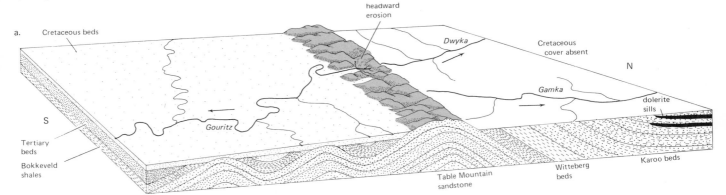

(a) Gouritz establishes its drainage pattern on Cretaceous beds and drains south from mountain range composed of Table Mountain Sandstone. Dwyka and Gamka drain northwards from the mountain range over slightly inclined surface.

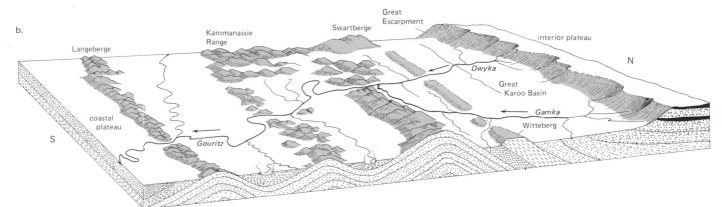

(b) Present situation: After considerable erosion, a new landscape has been exposed composed of east-west trending mountain ranges. The Dwyka and Gamka have been captured by the Gouritz, which has broken through the Swartberge, and they now drain southwards. The scarp of the interior plateau is gradually retreating due to headward erosion of Dwyka-Gamka drainage systems

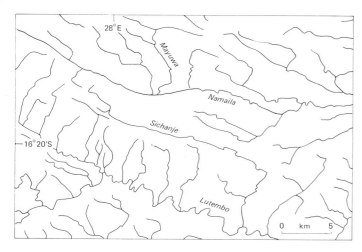

Fig 116. A rectangular drainage pattern in the Monze region of southern Zambia. The area is just north of the Batoka Gorge of the Zambesi River which also has a rectangular pattern caused by faulting

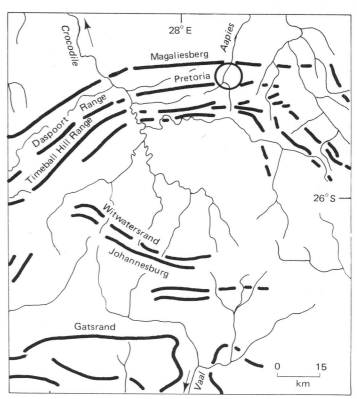

Fig 117. In the southern Transvaal, South Africa, the drainage patterns of the Crocodile and the Vaal have been superimposed across the east-west trending ranges

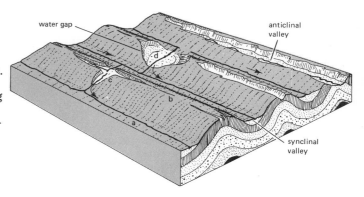

Fig 118. The influence of structure on drainage patterns in a folded mountain region. Some rivers flow along synclines (a), others have eroded valleys along the crests where tension has weakened the rock structure (b). Secondary streams have cut across the limbs of the folds (c) and eventually may cut right across the axes of the folds (d)

beds the river may reach lower, older formations of a different structure. If the river's pattern can be maintained across these older structures it is said to be superimposed. The middle Zambesi, for example, originally developed its valley on the surface of relatively level Karoo rocks but eroded downwards to a buried former landscape of mountain ranges lying transverse to the river's course. The river cut through these ridges in the gorges at Cabora Bassa, Lupata and Kariba. The Vaal River in the southern Transvaal also cut through thick beds of Karoo sedimentary rocks in the Parys region, eventually eroding down to the semicircular ridges which surround the eroded granitic core of the Vredefort Dome. The Vaal superimposed its original course, cutting through the ridges to the north and northwest of Parys. In the northern Transvaal the tributaries of the Limpopo—the Sand, Palala, Mogolakwena and Crocodile—are all superimposed on the Waterberge and Soutspansberge (Fig 117).

The synclines and anticlines of folded mountain regions (p. 82) strongly influence the river pattern, some rivers following the axes of synclines, others eroding valleys along the tops of the anticlines (Fig 118). A trellis pattern develops which is maintained even after considerable erosion has taken place. Eventually streams flowing down the sides of the anticlines to join the synclinal rivers erode water gaps and flow across the axes of anticlines. Gradually the surface layers of the anticlines will be removed and, if softer rocks lie beneath, thereafter erosion will be swift. In the more compacted rock of synclines, erosion will be a slower process and this differential erosion may produce inversion of relief, the former synclines standing as ridges above deeply worn former anticlines. An example of this is Table Mountain in the Cape Peninsula of South Africa (1087 m) which was once part of a syncline in folded sandstones. In general, the rivers of the folded Cape belt follow the synclinal valleys but several rivers have managed to cut across the anticlines. For example, faulting of the Olifants River Mountains formed the Tulbagh Gorge west of Tulbagh. A tributary of the great Berg extended its course by headward erosion through the gorge to capture the headwaters of the southeasterly flowing Breede to drain them westwards.

Fault alignment is a major factor in river pattern control in Africa. The upper Nile, for example, is strongly controlled by fault alignment, first flowing northeast from the L. Mobutu trough, then in a northwesterly direction, finally being directed in a northeasterly direction once more. The northwesterly flowing Aswa River north of L. Kyoga in Uganda is also aligned

along a faulted zone while in the Rift Valley of Kenya the Kerio valley is eroded along a fault line of the Kamasia horst. In Tanzania the course of the Pangani or Ruvu follows the floor of a graben at the foot of the Pare-Usumbara fault blocks and the Kilombero valley is similarly downfaulted against the Uchungwe-Usagara escarpment. The L. Rukwa graben, blocked at its northern end by lava, now provides a zone of centripetal drainage centred on fluctuating L. Rukwa.

In Nigeria, the course of the Benue is believed to be due to a major fault line and in Ghana the lower course of the Bonsa is controlled by a series of small faults. Similar fault control is also present in the lower Cunene valley and on the lower Lualaba.

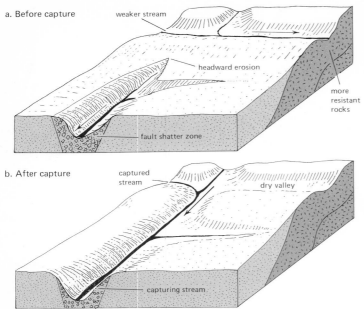

a. Before capture
weaker stream
headward erosion
more resistant rocks
fault shatter zone

b. After capture
captured stream
dry valley
capturing stream

Fig 119. Structural influences on river capture

River Capture

Rivers may extend their courses by headward erosion and by river capture. By these means a river course on more resistant rocks will be diverted to another drainage basin where erosion is easier. Headward erosion begins by the erosion of small scarps called headcuts at the starting point of a tributary. These headcuts will migrate backwards by surface wash flowing over the low scarp and eroding it and by undercutting and rotting of the scarp by seepage. The speed of this process, called headward extension, will be swifter on softer rocks or along zones of weakness such as a fault. Rejuvenation (p. 58) would also speed up the process. Thus, a strongly flowing tributary may eat through into a neighbouring valley and drain off or capture the more sluggish water of a weaker river. River capture may be recognised on a map in several ways. First, the capturing stream will be in a deeper valley than the captured stream; the latter often appears too small for its valley—it is a misfit stream—or it may dry up altogether. At the point of capture there is usually a pronounced change of direction

termed the elbow of capture. Between this elbow and the nearest headwater of the beheaded stream there is usually a dry valley called a windgap. Examples of river capture have already been referred to in the case of the Niger (p. 17), the Nile-L. Kyoga system, and the Great Berg-Breede. Other examples of river capture in Africa are numerous and only major examples will be discussed here. Alignment and headward extension along a fault-shatter zone are seen in the case of the Imo and Enyong rivers near Umuahia northeast of Port Harcourt in Nigeria (Fig 120). A tributary of the Otamiri, the Imo, has eroded a deep valley extending it northeastwards along a fault-shatter zone. The Imo cut through the Awka-Orlu uplands to the vale of the Enyong River and drained the headwaters of this less vigorous river into its own valley.

In Ghana, more evidence of capture is seen on the Volta (Fig 121). Originally the Black Volta and the White Volta were consequent rivers flowing to the Gulf of Guinea. A vigorous tributary of the White Volta captured the headwaters of the Black Volta and drained them eastwards. The Tano is thus the beheaded river and the dry gap is located to the west and south of Wenchi. A smaller example of capture in Ghana is that of the lower Bonsa south of Tarkwa which captured the headwaters of a river which formerly ran directly to the coast and drained them westwards.

Warping and tilting along the Zambesi-Congo watershed have produced a slight ponding back of the headwaters of the

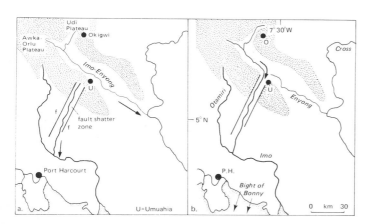

Fig 120. The capture of the headwaters of the Enyong River by the Imo, southeastern Nigeria

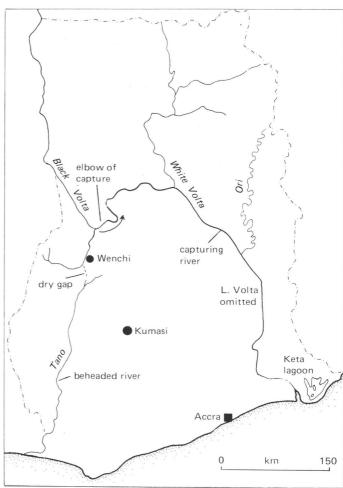

Fig 121. The capture of the headwaters of the Tano River by the Volta, Ghana

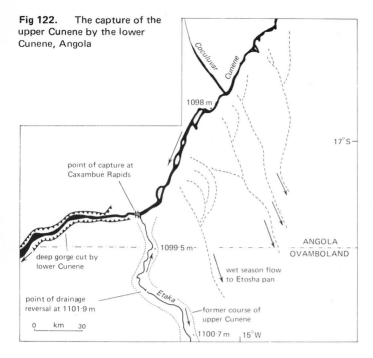

Fig 122. The capture of the upper Cunene by the lower Cunene, Angola

River patterns may be qualitatively described as trellised, dendritic or radial but this method does not provide a satisfactory means of comparing one drainage system with another on a statistical basis. Stream order classification (Fig 124) would seem to be the best means for quantative and comparative analysis of drainage systems.

Seepage and Springs

In the hydrological cycle not all water reaches the streams by ground surface routes. Where the rock is permeable or porous, rainwater seeps into the rock pores or runs down fissures until it reaches a hard pan layer or a band of impermeable rock. The water is then held in the rock pores and at lower levels the rock becomes saturated like a solid sponge. The surface of this saturated layer, the water-table, fluctuates considerably in the dry-wet climate of the tropics. During the rainy season the level of the water-table rises and springs may emerge from hill sides where the porous and impermeable rock strata meet. A line of springs may result which often becomes the site of several villages. A good example of a spring line lies along the northern edge of the Bandiagara sandstone scarp foot near the headwaters of the Volta River. With increasing dryness, the water-table falls and the line of springs may cease to flow leading to the abandonment of settlement. This has happened in the dry Hodh region north of the Manding Plateau near Bamako in Mali. The water-table may be, however, only a few metres below the ground surface especially in the beds of wadis (p. 112) and valley floors. But increasing dryness may cause the water-table to fall to considerable depths; in western Mali and southeastern Mauritania the level has fallen to 300 m below the ground surface.

Water seepage and springs are common in volcanic regions (p. 102). Rainfall caused by the higher relief of the volcano seeps along porous cindery layers in an intricate drainage network to emerge on the edges of the volcanic region. The Mzima Springs south of the Chuyulu line of ash cones in southeastern Kenya is a good example of this, the waters supplying Mombasa. The fault lines in the Danikil lowlands of Ethiopia are also marked by hot springs whose waters are full of chemicals from volcanoes.

Zambesi and may have diverted them into the Zaïre system. The indecisive drainage and swamps around L. Upemba and in the Lubudi valley near Kamina testify to the tilting. It is probable that the headwaters of the Kasai, called the Munhango in Angola, once flowed east then south to join the Zambesi system but were captured to drain southwards. The pronounced elbow of capture lies close to the point where the Benguela railway crosses the Angola-Zaïre frontier.

The Kafue River in Zambia displays an almost rectangular elbow of capture north of Nakala. The headwaters of the upper Kafue drained southwards along the now partially dry valley of the Machili but were diverted eastwards by the deeper incised lower Kafue to join the Zambesi near Chirundu.

In Angola the Cunene River provides a further example of capture. The upper Cunene, rising on the southern slopes of the Bihé Plateau, originally drained southwards to the Ekuma River which now flows intermittently into the Etosha Pan. The westward-flowing lower Cunene, rejuvenated by uplift, captured the upper Cunene and led its waters westwards to the Atlantic. A dry passage called the Etaka is now the only evidence of the Cunene's connection with the Etosha Pan (Fig 122). The Caculuvar, a major tributary of the Cunene, is also under threat of capture by the westward-flowing R. Coroca which has extended its headwaters through the Chela Mountains of the Great Escarpment.

The Pungwe-Nyakupinga river capture is seen in Fig 123. Another example in Zimbabwe is the capture of the Angwa River by the Hunyani.

The Ruaha River in Tanzania is a particularly active capturing river. By headward erosion it has cut back through the southern highlands via the Ngerengere Gorge to capture the drainage system of the Pawaga Plains and also to drain a huge former lake called Usangu. The drainage of the Usangu Plains was formerly westwards towards L. Malaŵi but this outlet was blocked by lava flows from the Rungwe volcano so that now the waters reach the Indian Ocean via the Ruaha and the Rufiji.

River patterns are the result of long periods of adjustment to structure involving capture, diversion, and perhaps reversal. It rarely, if ever, happens that the geomorphologist is presented with a pattern of consequent, subsequent and obsequent streams.

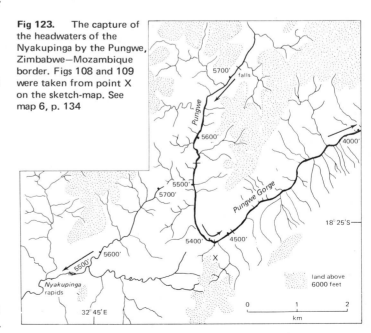

Fig 123. The capture of the headwaters of the Nyakupinga by the Pungwe, Zimbabwe—Mozambique border. Figs 108 and 109 were taken from point X on the sketch-map. See map 6, p. 134

Map Work

This work should be carried out on a small drainage basin within fairly easy reach of the educational centre so that map work can be followed up with fieldwork in the same region.

1. Drainage density is the sum of the lengths of the stream channels divided by the area of the basin. Thus if the lengths of all the stream channels add up to 120 km and the basin's area is 20 km^2, then the drainage intensity is 6. Analyse the drainage intensity of your basin.

2. Stream intensity is the number of streams in the basin divided by the basin's area. If the basin contains 24 different stream channels each linked with the others in a basin area of 20 km^2 then the stream intensity index is 1·2.

Work out the stream intensity and density for basins shown on your local survey maps. Calculate indices for the entire region. What is the value of such studies?

3. Stream order is a classification first suggested by R. Horton, and would seem to afford the best means for quantitative analysis and comparison of drainage basins. Streams in a basin form a hierarchy (Fig 124), first order streams being those which lack tributaries (1 in Fig 124); these unite to form second order streams (2), which in turn unite to form third order streams (3) and so on, until they combine to form the main stream which is ranked the highest order in the basin. The basin illustrated is thus a fourth order basin. Work out the stream order for your chosen drainage basin. Then for each stream make an analysis totalling the length of all first, second, etc., order streams, dividing them by their number to give mean lengths for each order and working out the mean gradient for each order. Do this for other basins in your region and write up your findings.

4. As a preliminary to fieldwork, obtain long profiles and cross sections of your basin's rivers noting carefully the position of irregularities which you can inspect in the field.

5. Calculating rainfall and run-off in a drainage basin: This work can be carried out on your local drainage basin if the necessary maps are obtainable from the meteorological department and the survey office.

Select two maps from your atlas which show a large drainage basin, e.g. the Congo/Zaïre and the zones of annual rainfall amount. Trace out the outline of the drainage basin by a line skirting the edges of first order rivers. Superimpose this traced outline on to the annual rainfall map. Trace the outlines of the different rainfall zones within the basin outline. Calculate the area of each zone in square kilometres using metric graph paper. Take the median value of each rainfall zone, (e.g. if it is 1500 mm–2000 mm the median is 1750 mm or 1·75 m) and multiply it by the area of that zone, then relate this to the river's average flow in cubic metres. Thus the calculation becomes:

$$\text{rainfall zone} \times \text{area of rainfall zone}$$
$$1750 \text{ mm or } 1\cdot75 \text{ m} \times 1500 \text{ km}^2 \text{ or } 1\,500\,000\,000\,000 \text{ m}^2$$
$$\text{or } 1500 \times 10^9$$

Make this calculation for each zone and find the total in cubic metres. This may come to 6 600 × 10^9. If the average mean flow of the river is 100 000 cubic metres per second this must be multiplied by 60 seconds by 60 minutes by 24 hours by 365 days to give the annual flow. This would be given as a percentage run-off figure:

$$\frac{3153 \times 10^9}{6600 \times 10^9} \times 100\% = \frac{3153}{6600} \times 100\% = 47\cdot7\%$$

Apply this exercise to the Congo/Zaïre Basin first, then the Niger, Zambesi or Orange and finally attempt it, if data is available, for a smaller basin in your own country. The Zambesi has a mean flow per annum of 31·4 million cubic metres, the Orange 7·6 thousand million, and the Congo/Zaïre 1·318 thousand million cubic metres per annum.

Fieldwork

1. Rainfall variation within a basin: One rain gauge will give a misleading impression of the rainfall of a whole basin. At least ten need to be set up at various points. It is not necessary to check them each day, and once a week or fortnight would be sufficient to give indications of variations. Simple rain gauges can be made by obtaining identical cylindrical bottles, filling them to just below the point where the bottle begins to narrow to the neck with old engine oil, inserting a red hot iron rod into the oil so that the bottle cracks cleanly, pouring away the oil, and inverting the bottle neck into the base. Mark the latter in millimetres from the base. Set your bottle rain gauges in strategic places, e.g. at the head of a first order stream, halfway down the course, near the base of a second order stream and so on. What percentage variation do you find in your various readings? Of what significance is this to your stream flow? Alter the position of your gauges from time to time. Write up your findings.

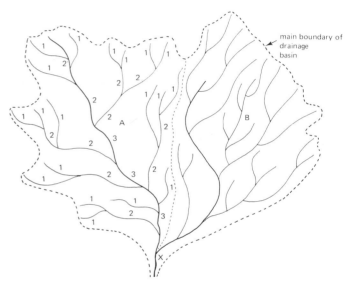

Fig 124. Stream order classification is shown in the sub-drainage basin A. Work out the order for sub-drainage basin B. What is the final order of the whole drainage basin at X?

2. Measuring groundwater levels: If there are wells or boreholes in your region it may be possible to obtain permission to test their depths from time to time by means of a float attached to a string or wire. Is the rise in water level coincident with rainfall or is it delayed? Explain. Does the nature of the surrounding rock have any effect? Write up your findings.

3. The shape of the river beds: Use a calibrated stick or ranging pole to sound the depths across the various sections of the river to obtain the shape of the bed. You should take advantage of the dry season to do this if the river ceases to flow. During the wet season the depth of water should be tested at the onset of the rains, from time to time during the season, and at the very end of the rains.

How will the shape affect the water velocity? Where will turbulence occur? Draw your cross-section to scale and write up your findings.

4. River velocities: The velocity of a river can be roughly obtained by means of floating corks or sticks in the current and timing them by stop watch between set points of known distance. A more accurate finding will be obtained by using a long cylindrical float of plastic weighted at one end so that it

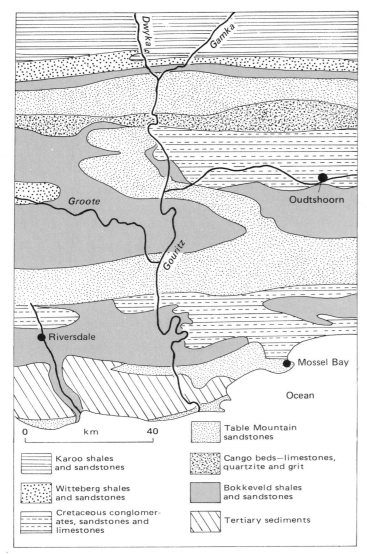

Fig 125. The geology of the Gouritz region shows the Gouritz also cuts at right-angles across the rock beds as well as across the relief trend. This map should be studied in conjunction with map 10 on page 138

Key:
- Table Mountain sandstones
- Karoo shales and sandstones
- Witteberg shales and sandstones
- Cretaceous conglomerates, sandstones and limestones
- Cango beds—limestones, quartzite and grit
- Bokkeveld shales and sandstones
- Tertiary sediments

0   km   40

stands upright in the current. This will be effected by lower currents as well. Compare the speed of a weighted and surface float. Calculate speeds in the centre and at the sides of the river. Where are the highest speeds and where the lowest? What relation have these to the shape of the river bed? Draw a sketch-map.

5. River loads—sediment: Immediately after heavy rains visit various parts of the stream and fill three or four narrow, tall bottles with samples of water taken from near the source, about half way down, and at a low point. Similar bottles may be filled with water from streams issuing from areas with different surfaces, e.g. pasture, woodland or cultivated zones. Allow these bottles to stand for several days and compare and make notes on the sedimentation you see.

6. Stream loads—bed deposits: Take samples from different sections of the river bed by using a scoop. Note the velocities of the stream at these points. Are the particles rounded or angular? Is the sediment well-sorted (the grains all of a similar size) or is it poorly sorted? What minerals seem to be most common in your samples? Explain your findings.

Where pebbles can be seen moving by saltation on the bed, select samples at random and measure their sizes. Relate your findings to the stream velocity at these points. Select stones which are not moving on the bed (choose at random) and examine their size and roundness. Is there a critical weight factor?

7. Meander measurement: Select a suitable part of the river on which there are several meanders and measure the wave-length, channel width, the radius of curvature of each meander, the degree of sinuosity, velocities, depths, and gradient of the stream. Check your data with the survey map and correct any inaccuracies on the map. Notice any riffles caused by unevenness in the bed of the stream. Measure the distance between these riffles and between the relatively quiet pools between them. What relationship have these statistics to the others you have obtained? Write up your findings with the aid of sketches and sketch-maps.

Questions

1. With reference to examples in Africa define the following terms:

stream volume or discharge; water table; springs.

2. Confining your answer to the continent of Africa and with the aid of sketch-maps and diagrams, describe the formation and morphology of the following riverine depositional features:

alluvial fans; deltas; sedimentary floodplains.

3. Define the term 'river profile'. What is the general nature of the profile of the major rivers of Africa? Illustrate with sketch-maps and diagrams.

4. Explain, with reference to examples in Africa, the meaning of the terms 'nick point waterfall', 'overflow waterfall', and 'outcrop waterfall'.

5. What have been the major effects on rivers in Africa of base-level change?

6. How have tectonic forces affected the drainage pattern of Africa?

7. Define, with reference to examples, the terms 'antecedent' and 'superimposed drainage'.

8. Describe the process of river capture. Illustrate your answer with examples of river capture in various parts of Africa.

9. Describe any fieldwork you have personally carried out in relation to any one small drainage basin. Use the headings:

Initial Preparation, Aims, Methods, Data Collected, Data Assessment.

The exact location of the drainage basin should be given.

10. What is meant by the term 'rejuvenation'? Give a reasoned account of the main landscape features resulting from rejuvenation in Africa.

11. What factors influence the cross-profiles of river valleys?

12. Examine the influence of rock-type on the long-profiles and cross profiles of river valleys.

13. How far do you consider it unsatisfactory to divide all river valleys into youthful, mature and old age sections? Refer to examples in Africa.

9 The Lakes of Africa

Africa's lakes may be classified according to their origins, and may be divided into major and minor groups. The major group includes:

(a) Depression lakes formed in downwarped basins which are either drained through gaps in the rims of the basins like L. Victoria, or, like L. Chad, are closed and have no outlet.

(b) Tectonic lakes caused by waters trapped in downfaulted valleys. Most of these lakes have formed since the Miocene period and include nearly all the lakes in the Rift Valley regions of East Africa.

(c) Volcanic lakes caused by blockage of a drainage system by lava flows, e.g. L. Amin (Edward) and L. Tana.

The smaller group includes:

(d) Crater lakes due to volcanic eruption (p. 100)

(e) Glacial lakes such as tarns and moraine-dammed lakes (pp. 75, 77).

(f) Seepage lakes caused by underground water rising to ground level, e.g. in oases and the chotts of the Maghreb.

(g) Coastal lakes and lagoons formed by enclosing sand-bars and sand-spits (p. 118).

(h) Pans caused by deflation in arid and semi-arid regions (p. 107).

(i) Oxbow lakes caused by river action and which are often temporary features (p. 49).

The major lakes of Africa have experienced wide fluctuations in their water levels and were once much larger. Between 30 000 and 8 000 years ago temperatures were some 5°C less than at present, evaporation rates were lower and lake levels higher. Lakes Chad, Rukwa and Rudolf, for example, were much larger about 9 000 to 10 000 years ago as shown by abandoned fossil lake shores or strand-lines and wide expanses of ancient lake sediments existing in today's arid regions (p. 111). Then, about 8 000 years ago, drier conditions set in and lake levels dropped. Some large lakes have disappeared entirely either because of evaporation or drainage. In the late Pleistocene period, for example, a large lake formed in the centre of the Congo basin and was later drained to the Atlantic by erosion of an overflow gorge in the basin rim (at the present Stanley Pool) and capture by a coastal river. Then some 10 000 to 15 000 years ago arid conditions prevailed with invasions of sand dunes in the south of the basin. Rivers were able to maintain their flow during these dry periods, fed by waters from the surrounding highlands.

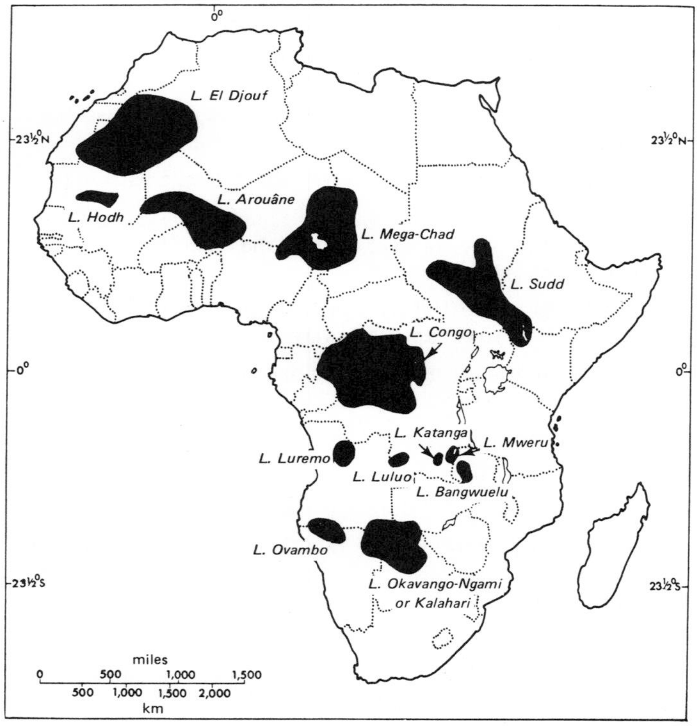

Fig 126.   Probable extent of Africa's major lakes before the beginning of the Miocene period, that is, before the formation of the Rift Valley

Depression Lakes

Lake Congo was a depression lake formed in a downwarp of the African surface. Lake Chad was formed in a similar way and now occupies the lowest western edge of the Chad Basin. Between 12 000 and 5400 years ago L. Chad expanded three times until it covered about 400 000 km^2, five times the present area of L. Victoria (Fig 128). The lake, called Mega-Chad, overflowed towards the Benue valley and so flowed via the Niger to the Bight of Benin. Then the lake shrank and was invaded by sand dunes which now give it its curious north-eastern coastline of drowned dune islands and miniature inlets. During the present century the lake's area has varied from 10 000 to 25 000 km^2 and in the 1960s it almost overflowed via the Bahr-el-Ghazel channel to the Bodélé depression. Lake Chad draws its waters from the Cameroon and Adamawa highlands and from the Bongo massif and the Central African Republic via the Logone. The lake remains fresh, probably due to seepage of salt-laden waters along the lake edges.

Lake Victoria also occupies a downwarped basin and, like L. Chad, is very shallow, being only about 80 m deep at most. Now covering 75 000 km^2, L. Victoria was once much larger and probably flowed northwestwards through a fault zone. Some 25 000 to 35 000 years ago this exit became blocked and the lake rose to 30 m above its present level. As the climate became drier the water surface dropped below the outlet level. Raised beaches are now seen at 18 m, 12 m and 3 m along the lake's shores. The ancient lake, called Lake Karunga, was finally drained northward about 3700 years ago by headward erosion and capture by a tributary of Lake Kyoga.

Fig 127. Tanzania: the southern part of Lake Tanganyika showing the steep dissected fault-scarp plunging to the lake surface. This is the eastern escarpment with the Kungwa Mountains in the background

Lake Chilwa in Malaŵi is another example of a depression lake and covers about 700 km². The lake's area, however, fluctuates with the seasons, the fringe marshes expanding and contracting. Again L. Chilwa was a much larger lake in pre-historic times and ancient beaches lie at 35 m, 24 m, 66·5 m and 12 m above present levels. At one time the lake drained northwards via L. Chiuta to the Ruvuma Basin and the Indian Ocean but now a wind-blown sand bar separates the two lakes.

Tectonic or Fault Lakes

The second group of large lakes, tectonic or fault lakes, are mainly associated with the Rift Valley (p. 90) although smaller examples are located outside this region. Lake Turkana lies in a shallow section of the Kenya Rift and is fed mainly by the Omo River, although the Turkwel from Mt Elgon supplies a seasonal flow and there are several seasonal springs along the lake shores. The lake was about 80 m higher 9500 years ago and the waters overflowed towards the Nile drainage system; subsequent lowering has exposed large areas of flat lacustrine plains. The water is saline but possible seepage keeps salinity lower than normal in this semi-arid zone of Kenya.

Lake Mobutu (Albert) is bounded by steep fault-scarps (p. 87) which rise 2000 m above the surface on the Zaïre side.

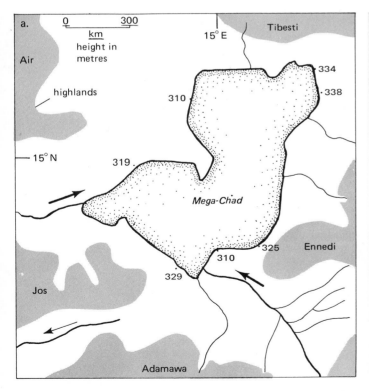

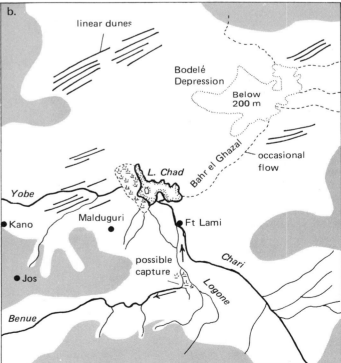

Fig 128. Evidence of climatic change is shown by fluctuations in the size of Lake Chad: (a) The lake about 10 000 years ago; during a pluvial phase, heavy rains swell the lake to ten times its present size. Mega, or giant, Chad was about the same size as today's Caspian Sea and extended some 300 km into northeastern Nigeria. (b) Lake Chad today is only 28 000 km² in area and is intensely saline; it contains approximately 20 million tonnes of dissolved salts.

69

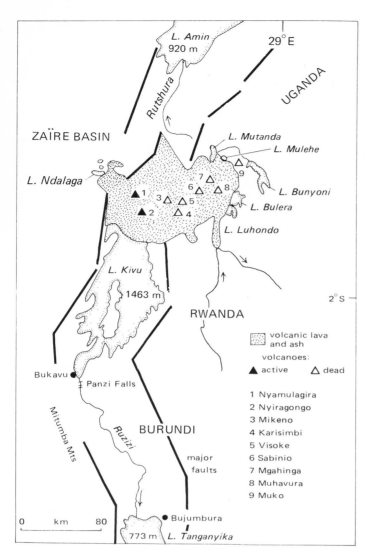

Fig 129. Lake Kivu is the largest of several lava-dammed lakes in the western arm of the Great Rift Valley. The lava was emitted by the nine Virunga volcanoes

Severe faulting caused the floor of the valley to sink to 1800 m below sea-level but was followed by a long period of deposition by rivers, the sediments reaching 2400 m in thickness, so that today the lake's maximum depth is only 58 m. The lake draws most of its water from the Semliki River and the supply from L. Victoria and Kyoga has little effect for it enters at the northern end of L. Mobutu near the Nile exit. This supply channel is a comparatively recent one caused by a river back-cutting to the Kabalega Falls.

Lake Amin (Edward) is also a typical rift lake with steep escarpments, especially on the Zaïre side, rising to 1200 m (Fig 129). The lake is fed by several rivers—the Nyamgarami, Ishasha, Rwindi and Rutshuru—and the waters leave via the Semliki to L. Mobutu. The lower Semliki eroded backwards over a volcanic sill to capture the northern tip of the lake. Lake Amin is connected to the shallow Lake George by the Kazinga Channel. Lake George is fresh water and lies in a very shallow basin being only 3 m deep. It is really an extension or bay of L. Amin.

Lake Tanganyika was formed when the western rifting during the Pliocene period interrupted the westward flow of the Malagarasi and Lukuga rivers. The Lukuga continued to flow westwards but the Malagarasi emptied its waters into the deep graben to form two lakes which gradually merged and then overflowed northwards to link with L. Kivu. The bed of the lake has two separate basins, the northern one 1310 m deep, the southern one 1470 m deep. The lake sides are steep and many collapsed stone blocks are visible beneath the surface waters. The through flow between the supply rivers Malagarasi and Ruzizi and the main outlet river, the Lukuga, ensure low salinity. Lake Tanganyika is noted for its stark, storm-swept shores.

Lake Malaŵi is the largest and deepest of the Rift Valley lakes being 770 m at its deepest (230 m below sea-level). The shoreline is not straight but runs in zig-zag fashion due to the intersection of two sets of faults which run NNW and NNE. In the north the Livingstone Mountains form a spectacular fault-scarp rising to nearly 3000 m above the lake level. The lake first appeared before the early Pleistocene and is older than the lakes further north. Its outlet via the Shiré River is often blocked by drifting sediments.

Lava-blocked Lakes

There are several lakes in Africa which may be described as lava-blocked lakes, for example, Lake Tana in Ethiopia and lakes Ndalaga, Mutanda, Mulehe, Bunyoni, Bulera and Luhondo in Western Uganda and neighbouring Zaïre. The largest example, however, is Lake Kivu formed by a blockage of the Rutshuru River by a late Pleistocene eruption of the Virunga volcanoes (Fig 129). The waters of the Rutshuru formerly flowed to L. Amin but were ponded back and flowed over an ancient dyke in the south, cutting a gorge and linking L. Kivu with L. Tanganyika. Thus L. Kivu does not display the straight shores of a rift lake but has a drowned ria-like coastline.

Lakes in Arid Regions

Large lakes once existed in the Sahara, Lake Arouâne being one of the largest (p. 17). Several thousand lakes existed in the pluvial periods between latitudes 14° and 22°N but most dried up about 7000 years ago leaving salt deposits, limestones and diatomite as, for example, at Bilma. These lakes were served by a network of rivers which today are dry or, at best, flow intermittently. Today much water from higher elevations escapes evaporation by flowing underground and collecting in vast subterranean reservoirs. Occasionally, such waters appear as springs to form rock pools and small ponds, e.g., the *guelta* of the Fezzan in western Libya. Here, the Wau en Namus, a volcanic cone, is surrounded by a saline lake in the middle of a sand dune zone (Fig 131).

Small Lakes

Some small lakes are formed by constructional or aggradational processes such as coral reef growth (p. 123) or sand-bar development. Lake Nabugabo on the western shores of L. Victoria is a small, shallow swampy lake covering about 20 km² and is no more than 5 m deep. It has been formed by a sand and shingle bar which has enclosed a small embayment of L. Victoria.

(Small lakes due to other processes are described on the following pages: volcanic crater lakes, p. 100; salt pans, p. 107; glacial lakes, p. 77; riverine lakes, p. 49.)

Fig 130. Uganda, Lake Mutanda, Kigezi Province: lava from the volcanoes Muhavura and Mgahinga in the background blocked the flow of a river to form this lake

Salinity of Lakes

The salt content of Africa's lakes may be considerably reduced by seepage of salt-laden waters from the lake by subterranean outlets and the constant inflow of fresh water in rivers. Some reduction of salt content may also be due to removal of salts by wind during temporary dry seasons when lake beds are exposed. This explains why lakes without surface outlets such as lakes Chad, Naivasha and Baringo are fresh water. Most lakes in Africa are, however, saline. The Karum salt lake in the Danikil depression of Ethiopia, for example, is composed of sea salt, for this region was once an arm of the Red Sea which was uplifted; it is also fed by intensely saline salt springs from Mt Dallol which have formed highly coloured and irregular landscapes of salt mounds and crags bordering the lake.

Fig 131. Southern Libya, Wau en Namus or Lake of Mosquitoes: these small lakes occur in a region of sand dunes and are formed by seepage around a volcanic ash cone

Map Work

Using a good atlas and metricated graph paper, produce a table of Africa's lakes showing:

Length   Breadth (maximum and minimum)   Total Area
Outlet Rivers   Inlet Rivers

In the table you should classify the lakes according to their formation and include man-made lakes (see p. 126).

Questions

1. With reference to specific examples, classify the lakes of Africa according to their mode of formation.

2. With the aid of sketch-maps and diagrams, describe the mode of formation of:

depression lakes;   rift valley lakes;   lava-blocked lakes;
crater lakes;   glacial lakes.

10 Glaciated Landscapes

Within the last million years the earth's surface has been invaded by immense ice sheets which, at their maximum extent, covered about 30 per cent of the land. During this glacial age, termed the Pleistocene, ice sheets advanced and retreated over northern Europe and North America. The Himalayas and the southern regions of South America, Australia and Africa, however, were buried under much earlier ice sheets of the Carboniferous period some 300 million years ago. Today only about ten per cent of the land surface or three per cent of the earth's total surface is under permanent ice, chiefly in the polar regions but also in the European Alps, central Norway and the New Zealand Alps. In Africa, active glaciers and ice caps, and their associated landforms occur on the giant fault block of the Ruwenzori Mountains in western Uganda, on Mt Kenya and on Kibo summit of Kilimanjaro in Tanzania. Many peri-glacial landforms also occur on the higher southward-facing slopes of the Drakensberg Mountains of Lesotho and Natal and on the high plateaux of Morocco and Algeria. These are landforms caused by extremely cold conditions bordering the main glaciated regions.

Snow Accumulation and Ice Formation

Snow forms where temperatures fall below freezing point and where there is considerable rainfall. The altitude at which snow forms varies with the seasons but there is a line, the snow-line, above which temperatures seldom, if ever, rise above freezing point. During colder seasons this line is at its lowest position. Here, where the accumulation of snow is greater than its rate of melting, permanent snowfields are formed. The height of the equatorial snow-line in Africa stands between 4500 m and 5600 m.

Fresh-fallen snow soon blows away or falls in avalanches to accumulate at the base of steep slopes. On more gentle slopes snow will gradually accumulate over the years until it forms an ice cap as on Kibo the summit of Kilimanjaro. New snow is composed of delicate crystals and trapped air and is white and fluffy. Gradually some air escapes, the crystals break down and the snow becomes more compact. The weight of further snow-falls, together with melting and refreezing causes the snow crystals to merge forming a coarser, stiffer snow called *firn* (Ger.) or *névé* (Fr.). Further compression squeezes out the remaining air and turns the firn into bluish-coloured glacier ice.

Glacier Flow

The pressure of overlying masses of snow and firn causes the ice beneath to move outwards from the centre of the ice sheet. In valleys, glaciers are also set in motion by the pull of gravity. The ice behaves somewhat like a brittle plastic when moving, the compressed, lower, harder layers moving slowly without melting while the upper layers tend to slide or sheer over them. Curved shearing planes called ogives are seen on thin glaciers. As the ice moves over or round rock obstructions or passes through narrow parts of a valley, tensions and pressures cause huge fissures or crevasses in the upper parts of the glacier. Multiple crevasses form at places where the ice drops over a ledge on its bed to form an ice-fall. The weathering and melting of the ice walls between crevasses creates irregular ice blocks called seracs. If the drop over a rock ledge is great an ice-fault may form.

The speed of the ice depends on the climatic season, the steepness of the underlying rock bed, and the rate of snow accumulation. It may vary from a few centimetres to 50 metres a day. Flow speeds are highest along the centre and in the upper parts of a glacier, decreasing towards the sides and bottom where the 'drag' of the ice on the rock surface occurs.

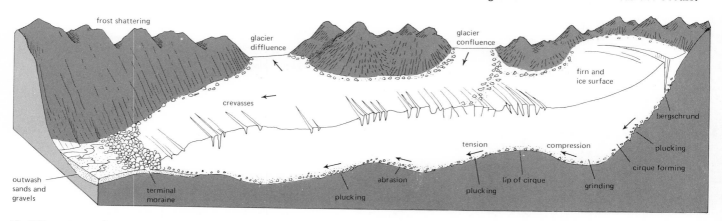

Fig 132. Some of the geomorphic processes associated with a valley glacier

Ablation

Glacier ice gradually moves to a lower point where higher temperatures cause melting. Some ice also evaporates. Pools form on the ice, retain the sun's heat and aid the melting process, reflected heat from valley walls and boulders on the ice also causes the ice to melt, and rivers and streams erode the ice. This ice loss is termed ablation. The level at which ablation exactly balances the firn accumulation is called the firn-line. If the rate of ablation is greater than the ice supply then the glacier will decay and retreat; if the two rates are equal, the snout will stay where it is, the glacier being in equilibrium; where the ice supply is greater, the glacier will slowly advance.

The glaciers in Africa have been retreating for some considerable time. The snow-line on the Ruwenzori fault block now extends down to 4750 m on the western slopes and to 4480 m on the east but there is evidence in the form of moraines that it probably once extended down to 3600 m. The glaciers which exist on the highest peaks were formerly much more extensive and originally reached down to 2000 m. Today, their fronts or snouts are generally above 4400 m but in some cases go down to 4335 m. Thus the glaciers are slowly retreating and some have entirely disappeared since the beginning of this century. In 1906 four glaciers were described on Mt Baker—Edward, Baker, Semper, and Moore or Mobuku—together with two large ice masses. Surveys fifty years later showed that the Semper glacier had entirely disappeared and had been replaced by a small snow-patch. The Semper glacier was 340 m wide and 180 m long in 1906 so that the ice loss is about 72 000 m² of ice in fifty years. Today the Mobuku glacier is the lowest ice form on the Ruwenzoris. During the late Ice Age, the glaciers on Mt Kenya and Kilimanjaro also extended some 1600 m lower than at present.

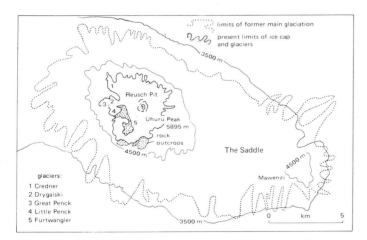

Fig 133. The present ice-cap on the summit of Kilimanjaro and the former limits of glaciation

The Nature of Ice Masses

Two major types of ice mass may be recognised. As we have noted, valley glaciers are located at high altitudes in tropical Africa. They are masses of ice which move down former river valleys from an ice cap or snow-field. Their movement and form is controlled by gravity, ice accumulation, and by the valley down which they flow.

Ice sheets (continental glaciers or ice caps) are broad ice domes, thickest at their centres and with semi-circular fringes.

Fig 134. Zaire, the foot of the Ruwenzori Glacier, western Ruwenzori Mountains: note the crevasses (A), supra-glacial till (B), glacial meltwater and erratics

Some 300 million years ago, during the Carboniferous age, extensive ice sheets invaded India, South America and Southern Africa. In South Africa traces of much older glaciations have also been discovered, the earliest occurring over 1000 million years ago. The region was not, however, affected by the later severe glaciation which occurred during the Pleistocene in Europe and North America although there were some periglacial effects (see below). Evidence of these ancient glaciations comes from the vast deposits of fossilised boulder clay or tillite in southern Africa. The ice sheets covered most of southern Africa (Fig 137) as the large ice sheets now cover Greenland and Antarctica. The tops of the higher mountain ranges such as the Drakensberg and the higher parts of Namibia must have projected above the ice surface like nunataks, the mountain peaks now standing above the Greenland ice cap surface.

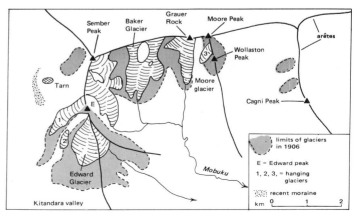

Fig 135. Present extent of the glaciers in the Ruwenzori region compared with their recorded extent in 1906

73

Four major ice sheets affected southern Africa and their paths of movement can be traced by scratches or striae on rock surfaces caused by stones embedded in the ice and tillite deposits. The four centres were located in what is now Namaqualand in Namibia, the northern Transvaal, Griqualand West, and a region east of the Natal coast now occupied by the Indian Ocean (Fig 137). The origin of the latter ice sheet is further evidence that the southern continents as a united whole were all covered by one huge ice sheet (p. 12).

Transport by Glaciers

Scattered boulders and other debris on top of the glacier are termed supraglacial moraine (Fig 138). As a glacier moves forward, pieces of rock prised from the sides of the valley by frost action and by general weathering fall on to the sides of the ice and are carried along as lateral moraine. Where a tributary glacier joins the main one, lateral moraines unite to form a medial moraine flowing at or near to the middle of the main glacier. Sometimes a tributary glacier is so narrow that it is compressed by the superior force of the main glacier and its lateral moraines are squeezed together to form a composite moraine. Where the glacier scrapes the side walls of the valley, rock particles become embedded in the ice and are transported as subglacial lateral moraine. Similar rock pieces work themselves down through the glacier to its base and here mingle with pieces plucked from the valley floor. This ground moraine or lodgement till embedded in the sole of the glacier increases the erosive power of the ice. The general term for all the moraine enclosed within the ice is englacial moraine. During warmer weather the top layers of ice may thaw and the englacial moraine just below the ice surface becomes mixed with supraglacial material to form a thick layer called ablation moraine.

The ice is not the only transporter of moraine. At the end of warm seasons when ablation is at its maximum, the surfaces of the lower parts of glaciers are crossed by scores of small streams. These carry off lighter debris as flow moraine either over the surface of the ice or down crevasses and holes called glacier mills or moulins. They join larger sub-glacial streams which emerge from the snout of the glacier as a milky-brown torrent and spread the moraine in a fan or kame. If the glacier snout is in retreat the kame becomes elongated up the valley in a recessional moraine; this is typical of some of the Ruwenzori glaciers.

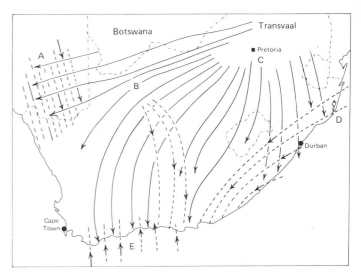

Fig 137. Directions taken by successive ice sheets in southern Africa:
(A) From Namaland, Namibia to south of Orange River
(B) A minor movement from northwest of Kimberley
(C) Large movement from Transvaal outwards to Natal, Namibia and Cape.
(D) From land area formerly lying northeast of present Natal coast
(E) From region to south of Cape Province coast

Glacial Erosion

Valley glaciers and ice sheets are powerful agents of erosion. Their sheer weight alone must weaken the surface of the rock beneath for even a thin glacier of about 150 m thickness creates a pressure of 15 kg cm^{-2}. As the ice proceeds it gradually works its way into joints, fissures and bedding planes in the rock surface. Blocks of rock and smaller pieces are prised loose, become embedded in the glacier sole and are carried away, a process termed plucking or quarrying (Fig 143). The remaining jagged surface is subject to further plucking.

The plucked rock pieces and loose debris embedded in the base and sides of the glacier abrade rock surfaces like a giant sandpaper or file. Rock is smoothed and ground to a fine rock flour and scarred by scratch marks called striations or striae.

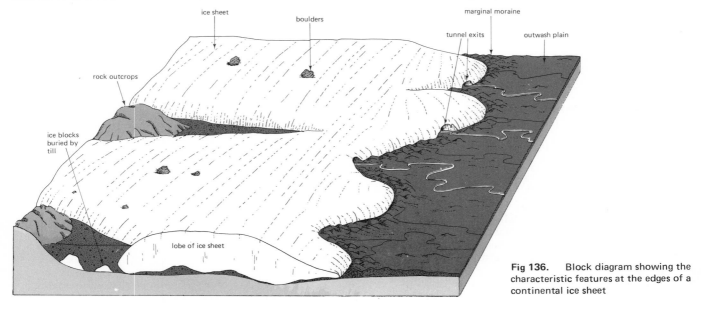

Fig 136. Block diagram showing the characteristic features at the edges of a continental ice sheet

Striations are seen on many rock surfaces in the Transvaal and Orange Free State where the covering deposits have been removed (Fig 140).

These erosive processes are not evenly distributed. The pressure on a band of rock is greater on the approach or stoss side, least on the lee side where plucking is more likely to occur. Again, the erosive power of the ice increases where the ice thickens, decreases where the valley widens and the ice thins as it spreads out.

Erosive Features of Glaciation

Fig 139 shows the chief features resulting from the glaciation of an upland region.

Cirques (corries, cwms or amphitheatres) are located at the head of the valley. Here snow, firn and ice once collected in hollows on the mountain side. The hollow was gradually deepened and widened by freeze-thaw shattering and by abrasion caused by the outward movement of the ice. These erosive processes are termed nivation. As the nivation hollow deepened, its back wall became steeper and more rugged due to plucking. The base of the hollow, where pressure of the ice is greatest, was subject to severe grinding. A steep rim-like threshold was formed and water often collects in the over-deepened part of the hollow to create a hollow-lake or tarn. Where several cirque glaciers united their erosive power was combined into one huge glacier. The result is a steep drop below the cirques called a headwall (Fig 141). Perched or hanging cirques occur on the sides of glaciated valleys where firn lodges in a small hollow. Hanging cirques are not as deep as head-of-valley cirques because their original glaciers were usually small and had less erosive power than the main glacier.

Horns or pyramid peaks develop on mountains which have been hollowed from several sides by cirque glaciers. Gradually as the cirque edges approached one another a thin, knife-like ridge, termed an arête, formed between them. These arêtes radiate outwards from the central peak and eventually the mountain becomes shaped like a pyramid.

Excellent examples of cirques separated by steep arêtes are found on Mt Kenya (p. 102) and most have tarns. Lacs Gris,

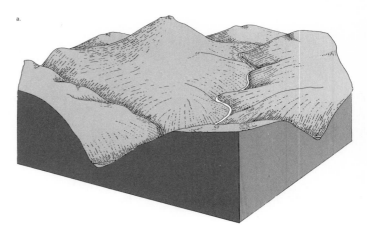

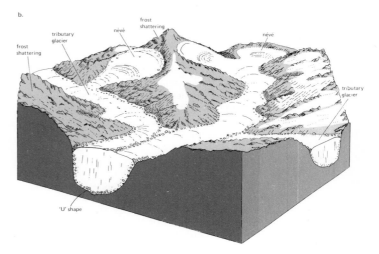

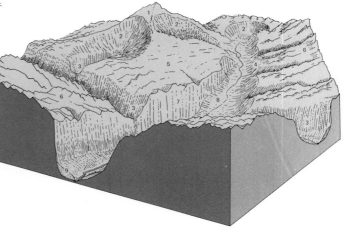

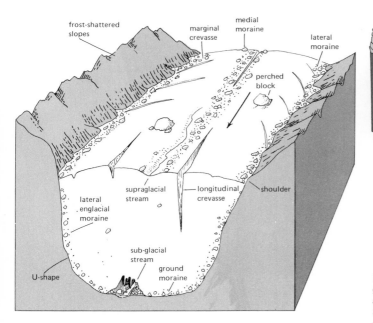

Fig 138. Block diagram of a section of a valley glacier

Fig 139. Glaciation of an upland area: (a) Upland region before glaciation. (b) The same region during glaciation. (c) The upland region after glaciation: 1. a pyramid peak; 2. cirques at head of valleys; 3. hanging valleys; 4. truncated spurs; 5. bench or alm; 6. rugged, frost-shattered uplands; 7. steep, almost vertical valley sides; 8. almost flat valley bottom; 9. lower slopes of fallen scree.

Fig 140.   South Africa, northern Cape Province: striae on rock surfaces and tillite of the Dwyka ice sheets at Nooigedacht, near Kimberley

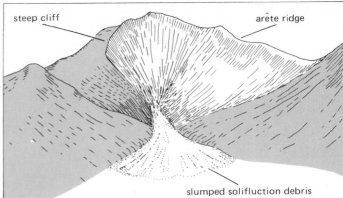

Fig 142.   Field sketch of the periglacial cirque and solifluction slump on Mount Currie, Natal, South Africa

Blanc and du Speke on the Zaïre side of the Ruwenzoris are cirque tarns and there are other hollows on the eastern side. Mt Margherita is similar to a pyramid peak and arêtes are formed by the sharp ridges of Speke and Gessi.

Arêtes, oversteepened glacial slopes, cirques and solifluction slumps also occur above 1650 m in the Drakensberg highlands of Lesotho. These features were formed under cooler conditions when temperatures were about 10°C lower than at present, probably about 3000 years ago. Heavy snowfalls and frost action occurred on the higher parts of the Drakensberg although ice-caps did not form. These features are therefore periglacial rather than true glacial forms; they have formed in the extremely cold zones on the fringes of the true glacial region. The nivation cirques of which there are about thirty, occur on the southward-facing slopes of the Witberge, the southern Natal Drakensberg, and the Central Range of Lesotho. One of the cirques is shown in the annotated sketch (Fig 142). Debris is removed by solifluction in which the

upper few centimetres of waste are moistened and slide under gravity over the frozen surface of debris beneath.

There are similar periglacial solifluction deposits 900 m thick in Algeria and 750 m thick in Morocco. These regions lay along the fringes of the major glaciations which affected Europe some 100 000 years ago and they probably experienced a tundra climate at high altitudes.

U-shaped valleys are formed when a valley glacier erodes both the sides and base of a former river valley by abrasion and plucking. Gradually, the original profile is changed to a broad or narrow U according to the rate of downward erosion. A noticeable feature is the abrupt change between the old angle of the river valley partly eroded by frost riving and the angle of the U, a feature called the shoulder (Fig 138). If the glacier filled the valley and the upper mountain slopes a characteristic bench is formed above the overdeepened valley. Glaciated valleys or troughs are usually straighter than the original river valley because the grinding and shearing action of the glacier reduces the faces of spurs to form truncated spurs and removes rocky outcrops..The smoothed sides of a glaciated valley often stand

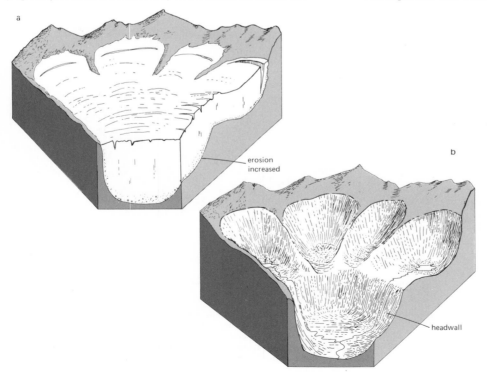

Fig 141.   Headwall formation by glaciers. (a) Several glaciers move down-valley from their névé fields. They merge at the point where they leave their cirques. The combined volume and weight of these smaller glaciers in one huge glacier increases erosion in the main valley. (b) The area after glaciation. The individual cirques are seen separated by sharp ridges or aretes. Where the glaciers joined is the headwall of the overdeepened main valley

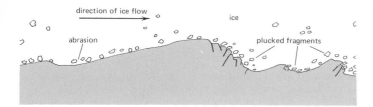

Fig 143. The formation of a roche moutonnée. The rock is plucked on the lee side and where the rock is jointed, smoothed on the approach or stoss side.

Fig 145. Uganda, Mount Baker, Ruwenzori Mountains: a large roche moutonnée photographed at 4 150m below the Mobuku Glacier. Note the smooth and plucked sides, the plucked valley sides in the background and the recessional moraine.

in strong contrast to the frost-shattered upper slopes. In the Ruwenzori region the Bujuku and the Mobuku valleys are U-shaped with steep vertical sides, flat floors and misfit streams giving poor drainage. Terminal and recessional moraines are features of these valleys as they are of those on Mt Kenya. Bigo is a glaciated trough end. In the Kaokoveld Highlands of Namibia, severe glaciation of granite and Table Mountain sandstone has resulted in several U-shaped valleys and striated surfaces.

Hanging valleys form in tributary river valleys which also filled with ice. These tributary glaciers joined the main glacier but, being smaller, their erosive power was weaker and consequently their valleys are shallower. Once the ice has disappeared their valleys appear high up on the sides of the main glacial trough. These hanging valleys are often drained by misfit streams which pour down into the main glaciated trough as hanging waterfalls or as wider bridal veils. Fig 135 shows several unnamed hanging glaciers west of Edward Peak and above the Mobuku glacier. In 1906 other hanging glaciers, now disappeared, were noted, some complete with bergschrund—the large crevasse at the head of a glacier caused by the pull of the ice away from the rock face.

The glaciated troughs of Mt Kenya and Ruwenzori also display irregular longitudinal profiles. The rock beds of the original river valleys are permanently altered by glacial erosion for, where the valleys narrowed the ice thickened, became heavier at that point, and eroded deeper; where the ice thinned out in wider parts of the valleys its erosive power decreased. The erosive action of the ice produces rock hollows and rock barriers (Fig 132) and, where the increased erosive power of the ice continues for some way down a valley it may form a distinct step. Later the hollows may be filled with melt- and stream-water to form lakes.

Where low outcrops of rock lie in the path of the ice the stoss or approach side bears the brunt of the ice pressure and is

Fig 144. Kenya, Mount Kenya: the twin peaks of Batian (5 199m) and Nelion (5 188m). The large Lewis Glacier is on the right to the left of Lenana Point (4 985m). The Tyndall, Heim and Farrel glaciers lie between Point Piggot and the central peaks. Mount Kenya is a frost-shattered volcanic plug. Note the aretes and pyramid peaks. See map exercise 7 on p. 145

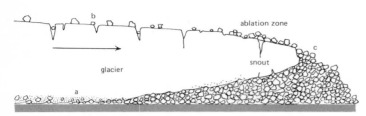

Fig 146. Debris deposition at the snout of a glacier. Lodgement till is being supplied at (a), angular frost-shattered rocks form supraglacial moraine at (b), to form the end moraine at (c).

ground to a smooth striated surface. As the glacier moves over the projection the lee side is subject to plucking, especially if the rock is well-jointed. A low rock form, smooth on the stoss side, jagged on the lee, is produced which indicates the direction of glacial flow. This is called a roche moutonnée (Fig 143). A prominent roche moutonnée exists below the snout of the Mobuku glacier (Fig 145).

A glacial feature which is partly erosional and partly depositional is the crag and tail form. If a large craggy outcrop such as a volcanic plug stands in the way of the ice, the glacier is forced to move round and over the crag. The lee side is protected and becomes a zone of deposition as ground moraine in the ice works loose. The stoss side is subject to tremendous pressure and abrasion and any collected moraine is carried over the crag to add to the depositional zone. The tail may also be composed of protected sediments of the former landscape missed by the ice.

Depositional Features of Glaciation

The moraine or till transported by the ice, sub-glacial streams and surface meltwater ranges from fine rock flour to huge boulders. As ice sheets retreated large stretches of the landscape were covered with this debris which is called glacial drift, till or tillite in its fossilised form. The most common deposit in the northern hemisphere is boulder clay, a matrix of clay containing all manner of small stones, pebbles and boulders of varying origins. In southern Africa, similar deposits called the Dwyka tillites are now seen as a bluish-green mudstone in which stones and boulders are embedded. The surface of these deposits—revealed by the stripping of later overburden—is level, undulating, hummocky or hilly. These tillite deposits reach thicknesses of between 350 m and 750 m in the extreme south of Africa but thin out to less than 30 m in the Transvaal. They outcrop over very wide areas as shown in Fig 140.

The rough texture of the tillites contrasts with the finer material washed out by meltwater streams at the edge of the ice sheet. Such fluvio-glacial deposits cover wide areas of the former outwash plain with gravels and coarse infertile sands in the Vryheid and Hartz River regions of the northern Cape Province of South Africa.

Large boulders, called erratics, are transported short or long distances by glaciers and ice sheets and are 'dropped' erratically when the ice melts. Boulders discovered in the Dwyka tillite along the Natal coast have been transported nearly 1500 km from their origins in the present northern Transvaal.

Besides these irregular deposits the Dwyka tillite contains several other characteristic features buried under later rock formations. Drumlins are a particular feature of the fringes of the ice sheets in the southern parts of Africa where the ice sheet began to thin out, deposition occurring where the ice thinned to about 60 m. The forward movement of ground moraine in the ice may be halted by a rocky obstruction and the elongated oval drumlin shapes built up. Long narrow tillite ridges akin to

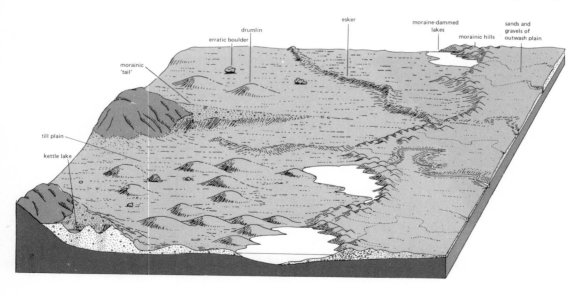

Fig 147. Glacial deposition landforms left after the retreat of an ice sheet (compare with Fig 136)

the eskers of the northern hemisphere formed in sub-glacial tunnels (Fig 147) and crag and tail features are also present in the tillite. A possible esker occurs in the Mobuku valley.

The glaciers on Ruwenzori and Mt Kenya have also left behind depositional moraines in their slow retreat. Moraines are carried to the snout of the glaciers in a continuous process and collect as terminal moraine in a crescent-shaped ridge. The deposits are particularly large if the glacier remained in equilibrium for a period; some of the moraines in the Mobuku valley are of this type. If the glacier retreats further and becomes stationary again a second, perhaps smaller, recessional moraine may be formed. Push moraines are formed from surface debris which has been pushed in front of the glacier ice front as if by a bulldozer. Evidence of push moraines is found at the lowest elevations of the East African glaciers, around 2000 m and the exposed tillite deposits in South Africa are partly of the push type.

Fieldwork

Glacial fieldwork is limited to bases in Kenya and Uganda within reach of Mt Kenya and the Ruwenzoris and to the periglacial regions of Natal and Lesotho. Soil samples below the lowest moraines, at the moraines, and above them can be taken to compare stone roundness, size and alignment. Distinctions should be made between fluvio-glacial and glacial deposits. Frost-rived scree can be examined and compared with the transported material. Mapping of moraines and other features not shown on the map should be attempted.

Questions

1. With special reference to the Mt Kenya and/or the Ruwenzori regions of Africa, describe the main erosion features of a glaciated upland.

2. Describe the main processes involved in snow accumulation and ice formation, glacier flow, and ablation. Relate your answer to glaciated regions in Africa.

3. With the aid of examples and sketch-maps describe and explain the mode of formation of the following glacial features:

striae; cirques; pyramid peaks; roche moutonnées; crag and tail; U-shaped valleys; hanging valleys; solifluction gravel.

4. Describe the main methods of debris transport by glaciers and ice sheets with particular reference to moraine transport in Africa.

5. Write an essay on 'The Dwyka Ice Sheets and their Effects'.

11 Tectonic Landscapes: Folding

Warping

When continuous lateral compression is applied to the earth's crust over a long period, the rocks react at first by gently warping. Few regions of Africa have escaped this warping process which has a decided influence on the present surface relief. This is particularly true of West Africa where the ancient shield of pre-Cambrian granites, gneisses, schists and quartzites of the Basement Complex has generally resisted the great lateral pressures which elsewhere have folded and uplifted the earth's crust. Here, uplifting and gentle warping of the basement complex have formed broad plateaux such as the Guinea Highlands and the Jos, Cameroon and Adamawa plateaux while the downwarped regions have become large shallow inland basins in which sediments eroded from the neighbouring plateaux have been deposited to form sandstones, limestones, clays and shales. The Sokoto and Chad basins and the Niger-Benue trough are sedimentary basins.

Downwarping of the East African plateau was also responsible for the formation of shallow L. Victoria (p. 68), the Tana Plains of Kenya—a downwarped region now filled with sediments—and the Wembere downwarp with its faulted margins in Tanzania. Downwarping and sediment infilling have also occurred in the L. Amboseli basin and in the upper Pangani river valley, and have caused the flat plains drained by tributaries of the Wami and Kilombero rivers.

Monoclines

Where rock strata dip slightly then level out again the feature is termed a monocline (Fig 148). In the present Natal region of South Africa a gigantic monocline began to develop in the early Jurassic period some 150 to 200 million years ago. The coastal regions were gradually depressed while the hinterland began to rise, and at the same time the broad axis of the monocline began tilting or pitching slightly towards the south. Today, the Natal or Lebombo monocline can be traced for about 320 km in the Karoo rock system which runs almost horizontally from the Drakensberg Mountains to Pietermaritzburg and then dips gently towards the Indian Ocean.

Monoclinal downwarping has resulted in the partial drowning of the coast of West Africa, the sea invading the former deep river valleys to produce estuaries and rias in Senegal and Sierra Leone and at the mouth of the Cross River in Nigeria. Where deep river mouths were absent, the coast was not indented by monoclinal sinking as shown by the relatively straight coasts of Natal and Angola.

Domes and Basins

When pressure on rock beds is exerted laterally from several angles simultaneously or from beneath, the strata may rise to form a dome. Some 12 to 25 million years ago crustal movements in Kenya led to the formation of a broad low dome about 300 m above sea level centred on the present Nakuru region. The doming was accompanied by the monoclinal submergence of the coastal regions below the sea. During this period a radial drainage pattern developed over the surface of the dome with rivers draining to the Congo basin, the Indian Ocean and the Nile (Fig 112). Further uplift occurring about 9 million years ago during the Pliocene increased the height of the Kenya dome to approximately 1300 m. Similar doming occurred in Ethiopia and the large downwarped depression between the Kenyan and Ethiopian domes formed the present infilled sedimentary plains (the Tana Plains) between Wajir and Garissa and in southern Somalia.

During the late Tertiary period the southern part of Africa underwent doming which raised the present central highveld section and caused the upwarping of regions such as Inyanga in Zimbabwe to over 2100 m.

Similar doming on a smaller scale has occurred in the Parys-Vredefort region of southern Transvaal province. Here several beds of sedimentary rock (shale, quartzites, limestones) dip steeply outwards and away from an ancient granite core to form the Vredefort Dome. The upper part of the dome has been eroded but the surrounding sedimentaries outcrop to form a ring of low escarpments some 40 km in diameter around the central core.

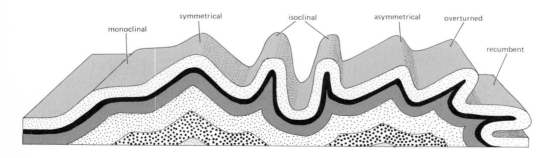

Fig 148.   The chief types of folds

80

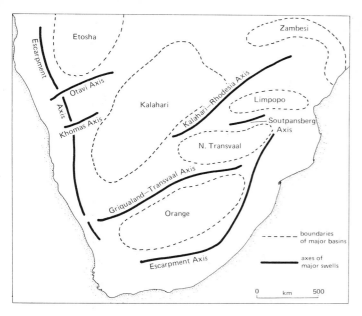

Fig 149.   The 'basin and swell' nature of the surface of southern Africa

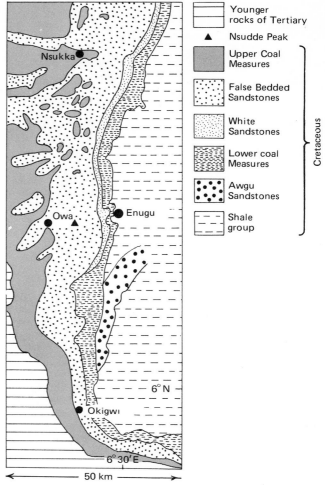

Legend:
- Younger rocks of Tertiary
- ▲ Nsudde Peak
- Upper Coal Measures
- False Bedded Sandstones
- White Sandstones
- Lower coal Measures
- Awgu Sandstones
- Shale group

(Cretaceous)

Fig 151.   Geology of the Okigwi–Nsukka region. Compare with the cross-section on the block diagram in Fig 150

In some cases pressure from all sides produces a broad shallow basin instead of a dome. On a very large scale the Kalahari Basin is composed of strata which dip gently inwards but the basin has been buried at the centre by thick sand deposits. A smaller example is the swampy Chilwa-Chiuta basin in southern Malaŵi where the shallow downwarp has been filled to great depths with sediments. Other downwarped basins include those occupied by lakes Victoria, Bangwuelu in Zambia, Chad in Nigeria, and the Makarikari swamps of Botswana (p. 68).

If lateral pressure is steadily increased the rock strata begin to fold. Slight compression forms arches and troughs (anticlines and synclines), the section linking these being termed a limb. Erosion of the limb of an anticline may produce a prominent erosion scarp or cuesta and it is in this way that the Okigwi cuesta in Nigeria was formed (Fig 150).

Under extreme pressure the anticlines begin to rise and the synclines become relatively deeper. If the pressure is uneven or the strata are squeezed so tightly together that they cannot stand unsupported, the limbs will assume a roughly horizontal position nearly parallel to the ground and form a recumbent fold (Fig 148). In some cases large anticlines and synclines may

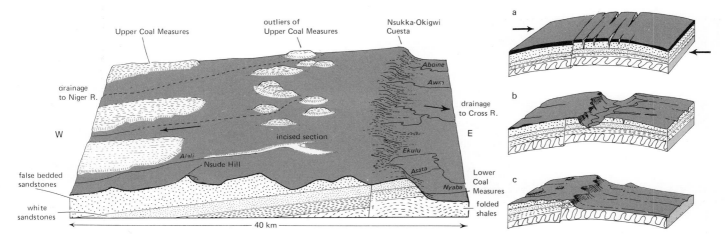

Fig 150.   Nigeria: the Okigwi–Nsukka cuesta. Left: Block diagram showing the geological structure and its influence on the surface relief. Right: Block diagrams illustrating the formation of the cuesta.

(a) Lateral pressure causes upwarping of rock beds; (b) Crest of upwarp is eroded first. (c) Eastern limb removed.

81

Fig 152.   South Africa: the intensely folded strata of part of the Zwarteberge, near Prince Alfred's Town, Cape Province, The bushes at the base of the mountain give an indication of scale.

themselves become wrinkled into smaller folds to form an anticlinorium or a synclinorium. This has happened in the Swartberg and Langeberg Mountains of the southern Cape folded belt. During such folding the rocks may become highly deformed and be stretched out up to thirty times their original length.

Folding on a regional scale (orogenesis) creates whole chains of mountains and occurs where a continental plate comes into contact with the moving oceanic crust which then sinks below the continental plate (Fig 153). Some orogenic beds have formed along the sites of long sedimentary basins or geosynclines at the edge or between continental plates. Vast quantities of sediments eroded from the surrounding continental highlands are deposited in these geosynclines causing the base to sink under the accumulating weight. In some fold mountains the thickness of the sedimentary strata reaches 15 km indicating the vast depth of the original syncline. As the

geosyncline deepened, the crustal rocks weakened and stretched and igneous intrusions occurred. The continental plates converged, squeezing the sedimentary rocks into folds which yield to further pressure by faulting. The formation of fold belts may take several hundred million years or only a few million years.

Orogenic Periods in Africa

Africa has experienced at least seven major mountain building periods or periods of orogenesis (Fig 1). The oldest orogenesis occurred more than 2500 million years ago affecting the eastern half of southern Africa. Following this another period of mountain building affected Liberia, Sierra Leone and the Ivory Coast region, parts of the Cameroon and the Kasai region of the Congo basin. Further folding affected much of Africa west of longitude 0° and most of the Congo basin. All these regions have been eroded and have remained undisturbed for some 1500 million years. Such regions are called cratons.

Mountain building occurred again between 200 and 1110 million years ago in Namaqualand, Natal and in central Kenya while the fifth orogeny of 100 to 550 million years ago affected much of the rest of Africa. The Cape folded belt of South Africa formed during the middle Palaeozoic and early Mesozoic periods. Finally the Alpine orogeny some 12 to 50 million years ago created the Atlas ranges. These younger folded regions are termed orogens.

The Ghana/Togo Folded Ranges

The older folded belts have long since been removed by erosion or have been buried beneath thick sedimentary deposits. For example, most of the rocks in Ghana show traces of intense folding but this has occurred so long ago during the pre-Cambrian and Palaeozoic eras that the former mountains no longer exist. The Akwapin-Togo-Atakora Mountains, however, have survived as a range composed of very closely folded strata with peaks rising to 425 m in Ghana and to 850 m in Togo. The ranges trend southwest to northeast in Ghana and Togo and in some cases the mountain cores of resistant rock stand up as sharp serrated ridges (Map 9, p. 137).

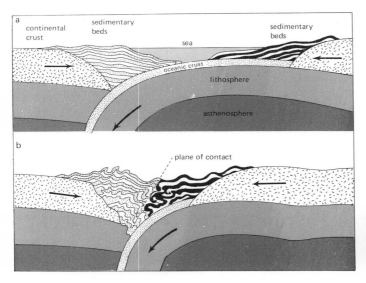

Fig 153.   Fold mountains formed by continental movement: (a) Lithosphere plate moving laterally is forced into asthenosphere on contacting another lithosphere plate. Continents move towards one another pushing sediments derived from their surfaces before them. (b) Sedimentary beds collide as continents move closer and are crumpled into folds.

Fig 154.   South Africa: Table Mountain, Cape Town, viewed from the eastern landward side. The mountain is composed of sandstone and is the remnant of a syncline.

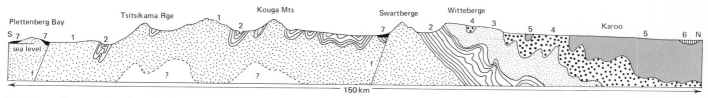

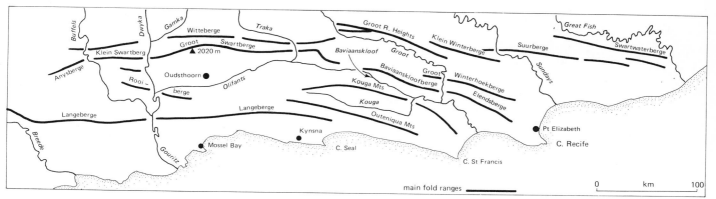

Fig 155 Cross-section through the southern Cape Fold belt. Key to numbers: 1. Table Mountain Sandstones; 2. Bokkeveld series (shales and sandstones); 3. Witteberg Series (quartzites, sandstones and shales); 4. Dwyka Series (chiefly tillites of glacial periods); 5. Ecca Series (grits, sandstones, shales); 6. Beaufort Series (sandstones, shales, mudstones); 7. Cretaceous beds (conglomerates, sandstones and clays)

Fig 156. The South and South-East Cape Province: trends of major fold mountains

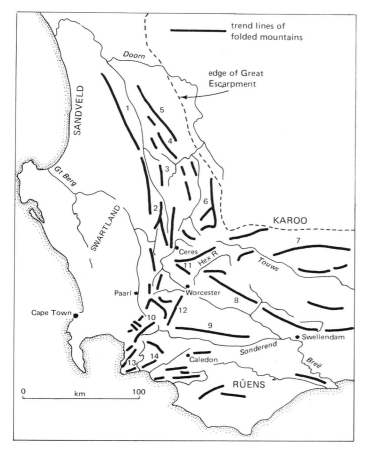

Fig 157. The South-West Cape Province: Trends of major fold mountains. For key to numbers see text

The Cape Folded Ranges

In the southern Cape region the folding of Table Mountain Sandstones, formed in a geosyncline between the plates of Africa and Antarctica, occurred during the late Triassic and Jurassic periods but the actual causes of this folding are not yet clear. The folded belt can be sub-divided into two regions:

(i) *The southwest Cape region* (Fig 157) in which two series of ranges trending south–north and west–east meet in a complicated knot of ranges in the Worcester and Ceres districts. The folds are composed of quartzites, shales and sandstones. The major north-south trending ranges are the Olifants River Mountains (1 on map), the Great Winterhoek Mountains (2), the Koudbokkeveld Range (3), the Skurberg (4), the Cedarbergen (5), and the Swartruggens (6), all mostly above 1800 m. The main eastward-trending ranges are the Witteberge (7), the Langeberge (8), and the Riviersonderendeberge (9). In the complicated group the ranges lie almost at right angles to the other range trends, for example in the Hottentot Holland Mountains (10), the Hex River Mountains (11), and the Stettynberge (12). Two of the ranges, the Kogeberg and the Paardeberg reach the sea and form steep cliffs at C. Hangklip. Table Mountain is a remnant of a downwarped fold. Several basins bounded by faults are found in this region, chief of which are the Elgin, Tulbagh, Worcester (Fig 166), and Robertson-Ashton basins.

(ii) *The southeast Cape region* (Fig 156) has lower east–west trending ranges which end at Cape St Francis and Cape Recife. The highest parts are located in the Suurberge. The northern ranges, the Suurberge and the Klein Winterberge, are composed of quartzite while the southern ranges—the Groot Winterhoekberge, the Baviaanskloofberge, the Kouga Mountains, and the Tsitsikammaberge are of Table Mountain Sandstone.

83

Fig 158. South Africa, Worcester area, South-west Cape Province, north-east of Cape Town: the level summits of the Gondwana erosion cycle (background); the lowlands are the advancing African erosion surface

Fig 159. Diagrammatic cross-section through the Chimanimani Mountains, eastern Zimbabwe showing an overthrust fold with sheets of rocks or nappes being thrust forward along thrust planes

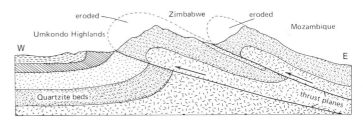

Fig 160. Zimbabwe: this region, south of the Kariba township, lies between the stable cratons of central Zimbabwe and the Zambia—Tanzania cratons to the north. The rocks between these two stable regions were subject to squeezing while they were still in a plastic state and became twisted and folded. The trends of the twisted fold ranges are easily seen in this vertical air photograph at a scale of approximately 1:70 000. Locate evidence of superimposition and structural control by syncline on the drainage pattern

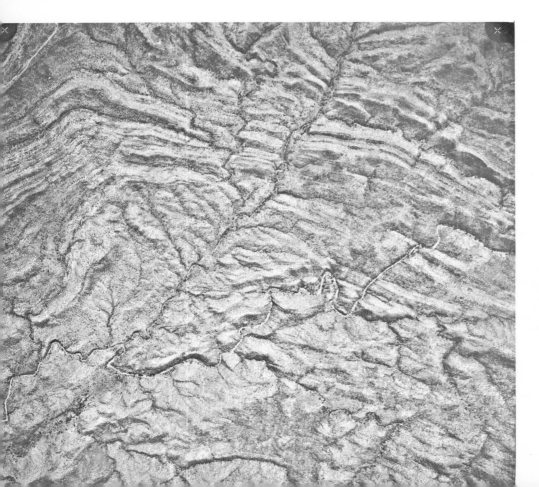

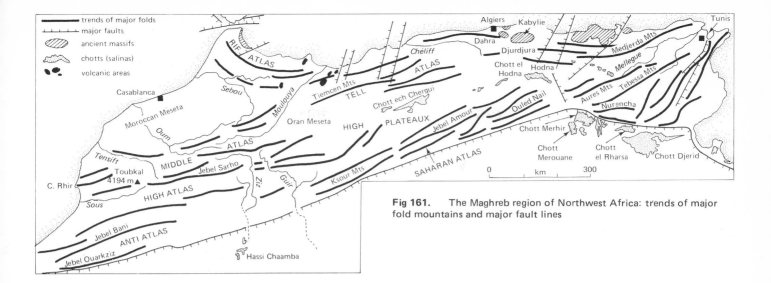

Fig 161.   The Maghreb region of Northwest Africa: trends of major fold mountains and major fault lines

The Atlas Folded Ranges

The Atlas Mountains of the Maghreb are structurally similar to the great alpine system of Europe being formed during the same geological period when the African and European plates moved towards each other. During the folding, plutonic igneous rocks were intruded into the crumpled sedimentaries of limestone and sandstone. Again, there are several distinct groups of ranges (Fig 161):

(i) *The Coastal Ranges* which rise to over 2100 m extend westwards into the Tell Atlas which ranges between 1000 m and 1800 m, its highest parts lying in the Great and Little Kablye Mountains. Here again the ranges end as cliffs at C. Blanc in Tunisia.

(ii) *The High Atlas* stretches some 800 km from the coast of Morocco near Agadir in a northeasterly direction. There has been much crystalline igneous intrusion here producing high peaks such as Jebel Toubkal (4194 m), Irhil M'Goun (4071 m), and Volcan du Siroua (3306 m).

(iii) *The Saharan Atlas* consists of a series of limestone and sandstone ridges trending northeastwards—the Monts des Ksour, the Djebel Amour and the Monts des Ouled Naïl, the latter extending into the Aures Mountains which rise to between 1200 m and 1800 m and end in C. Bon.

In the west lies the *Moroccan meseta* or tableland which acted as a huge block against which several of the fold ranges were crumpled. The tremendous pressures caused the southern parts of this plateau to subside.

Questions

1. What have been the effects on the surface morphology of Africa of warping, doming and folding?

2. Describe the forces which are responsible for the creation of folded mountain belts in Africa. Choose any two folded regions and write a brief description of them.

3. Describe and explain the mode of formation of the following features due to orogenesis, illustrating your answer with examples drawn from Africa:

upwarp; downwarp; monocline; anticline; syncline; erosion cuesta; anticlinorium.

Fig 162.   Morocco, the western part of the High Atlas: the river has cut a deep gorge through almost horizontal strata

85

12 Tectonic Landscapes: Faulting

Joints and Faults

Pressures and stresses in the earth's crust may be so intense that they cause the rock strata to fracture into large cracks called joints. Joints are most numerous in the upper parts of the earth's crust, becoming smaller and less frequent with depth until at about 20 km, where rocks are more plastic, fractures rarely occur (compare rock joints with mud cracks and ice crevasses). When severe tension or pressure causes movement of rock faces against each other along a joint, rock beds are displaced to produce a fault (Fig 163). The angle of displacement of faulted rock beds depends on the original angle of the joint. The resulting plane of the break or fault plane stands at an angle to the vertical, the angle being termed the hade. The amount of vertical displacement is called the throw and the horizontal displacement is termed the heave (Fig 164).

Pressure, tension, and the force of gravity operate in varying degrees to produce several types of fault in which the amount of displacement may be slight or may extend over many kilometres depending on the force involved (Fig 165). A normal or tension fault is caused by the stretching of the rock crust. This is the most common type of fault, rocks being very much weaker and prone to faulting when under tension than rocks under compression which bend or fold before fracturing. The hade of normal faults is usually between 30° and 45° and the throw may be tremendous. The Worcester Fault in South Africa (Fig 166) which extends for 370 km from Piquetberg to near Riversdale has a throw of approximately 3350 m in the Worcester region and a very steep hade. Throws exceeding 3000 m have also been recorded in the L. Mobutu trough in Uganda. A reversed fault is caused by compressional forces, one block of rock overriding another and sometimes depressing it, while a tear fault is due to predominantly horizontal movements. Low angle compression faults causing one sheet of rock to override another are also termed thrust faults. The grinding movement of one fault face against another often creates a striated or smooth fault face, a process known as slickensiding. If two parallel faults lie close together the rock between them may become crushed to form a fault line filled with shattered rock fragments termed fault breccia.

Most faults rarely occur as single isolated features but are usually found in parallel series. Sometimes tension allows each

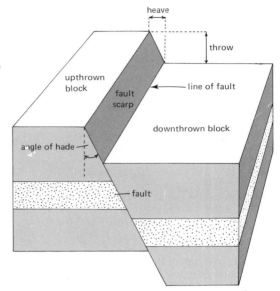

Fig 164.   The chief terms used in describing faults

rock section between faults to slip down relative to their neighbours to produce step faults (Fig 163). The eastern scarp face of the Great Rift Valley west and southwest of Nairobi in Kenya descends to the Kedong Valley and the Nkama Plains in a series of such step faults. In Malaŵi the western side of the Nyasa trough displays low crystalline ridges about 50 km long due to step faulting and is in marked contrast to the steep down throw on the eastern side. Sometimes faults do not fully develop and fault splinters result (Fig 163) producing low escarpments which run for a few hundred metres and then peter out. Further complications occur when the initial fault is followed by a further tectonic movement to produce a double fault. Moreover, in such regions of crustal weakness, faulting and jointing provide easy escape routes for upwelling magma. Both double faulting and the masking effects of lava flows are illustrated in the Nguruman Escarpment which lies west of L. Magadi in Kenya (Fig 167). Similar double faulting has occurred in the Nyasa trough of Malaŵi.

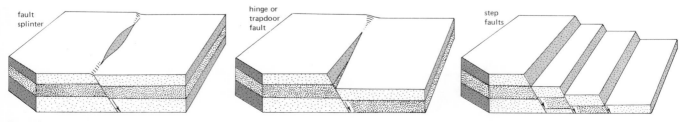

Fig 163.   Some landforms associated with faulting

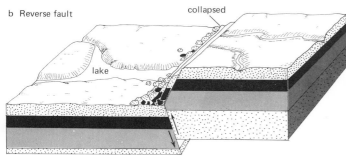

a Normal fault

upthrown block

downthrown block

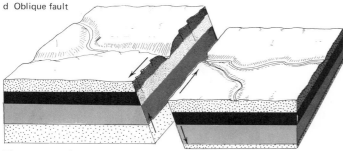

b Reverse fault

collapsed

lake

The right-hand block has moved relatively
upward overriding the left-hand block

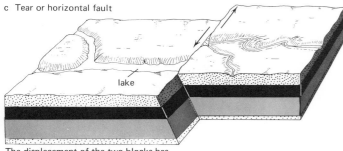

c Tear or horizontal fault

lake

The displacement of the two blocks has
been horizontal with no vertical movement

d Oblique fault

The movement has been both horizontal and
vertical

Fig 165. The major types of faults and associated surface features

Fault and Fault-line Scarps

The initial landform produced by a fault is the sharp step-like
fault-scarp which may run for hundreds of kilometres along the
fault line, e.g. the Worcester Fault. On young fault scarps where
little erosion has taken place the height of the escarpment
indicates the amount of downthrow. The young fault-scarp
along the western shores of L. Mobutu extends for some 250 km
and has a scarp face over 300 m high (Fig 171). River erosion of
the scarp face and deposition of debris are here in their early
stages. The Livingstone fault scarp which rises 1500 m above
L. Malaŵi's surface is of greater age and has been deeply
dissected into rugged V-shaped valleys and sharp ridges.

Eventually the fault-scarp will be removed by erosion leaving
an abrupt change of rock type on the levelled surface along the
line of the fault. If the rocks on one side of the fracture are
more resistant than those on the other differential erosion will
form a fault-line scarp (Fig 168). Sometimes the eroded fault-
line scarp will face the same direction as the original fault-scarp
to form a consequent fault-line scarp as in the case of the
Nguruman Escarpment. But if the strata of the downthrown
block are the more resistant the scarp will face the opposite
direction as an obsequent fault-line scarp (Fig 168). The Inanda
Mountains in Natal, South Africa, display an obsequent fault-
line scarp of Table Mountain Sandstone overlooking a vale of
softer rocks.

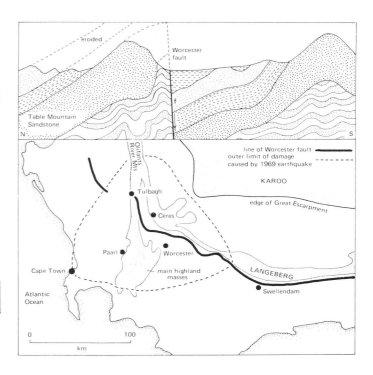

Fig 166. The Worcester Fault, Southwest Cape Province, South Africa.
Above: Diagrammatic cross-section (vertical scale greatly exaggerated)
to show the relationship of the rock beds on either side of the fault.
Below: Map of the region showing the position of the fault in relation
to the folded ranges

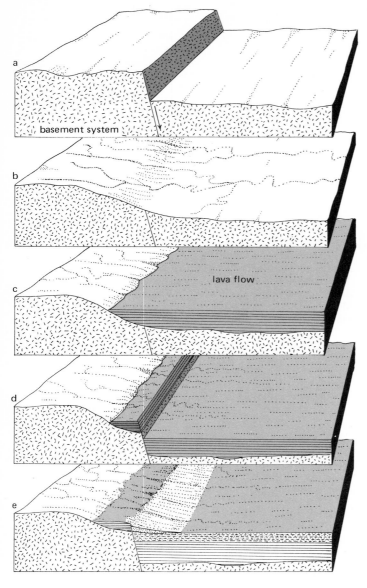

Fig 167. The Nguruman Escarpment, west of Lake Magadi, Kenya: (a) first normal fault; (b) erosion; (c) burying by lava flow; (d) second normal fault on old fault line; (e) erosion to present surface

basement system

lava flow

Grid Faulting, Horsts and Graben

Where lines of faults cross a pattern termed grid faulting is created. Such grid faulting is most noticeable from the air since river erosion along the fault lines often produces a rectangular drainage network. The Shiré Highlands region in Malaŵi displays this cross faulting. Pressures and stresses in such regions create a landscape of fault-bounded mountain blocks or horsts separated from each other by suppressed blocks called graben (Fig 169.) Originally such landscapes must have been sharply angular but long periods of erosion deeply etch the blocks and deposit waste material to fill in the basins.

One of the finest examples of a horst is that of the Ruwenzori Range which lies within the western arm of the Great Rift

Valley (Fig 170). Here upwarping of the former surface has reached 4500 m to create a huge tilt block, the highest point of which is Margherita Peak on Mt Stanley (5109 m). The western facing fault-scarp, extending 105 km from southwest to northeast, descends over 4000 m to the floor of the Semliki valley. Eastwards the block tilts into the L. George depression without any fault occurring until the eastern shores of the lake. The greatest width of the Ruwenzori block is about 45 km but the northern end of the massif narrows between two fault scarps to about 12 km. Here the block dips northwards and gradually disappears below the sediments of the L. Mobutu trough.

Numerous other smaller but still impressive horsts occur in East Africa—the Nyiru, Ndoto and Matthews ranges between Mt Kenya and L. Turkana; the Usumbara and Pare Mountains, the Iramba Plateau, the Uluguru and Mahenge blocks, the Mbeya Mountains and the Kungwe and Ufipa plateaux of Tanzania (Fig 175). The Ufipa Plateau is a large horst which is tilted towards the northeast to a maximum height of 2418 m and here its fault-line scarp dominates the southwestern edge of the L. Rukwa trough, a typical downthrust graben. The Usumbara fault-scarp in northeastern Tanzania rises to over 1500 m above the surrounding surface. The Mbeya Mountains show evidence of two stages of uplift and erosion since the block surface at about 2400 m above the surrounding plain gives way to a lower level surface at about 1200 m.

Other examples of block mountains include the Great Kharas Mountains of Namibia which rise to 2201 m in the Schroffenstein and the Sijarira Range south of L. Kariba in Zimbabwe (Figs 172 and 173).

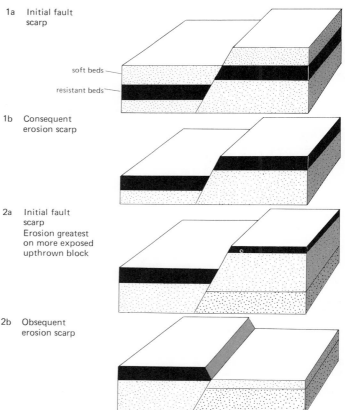

1a Initial fault scarp

soft beds

resistant beds

1b Consequent erosion scarp

2a Initial fault scarp
Erosion greatest on more exposed upthrown block

2b Obsequent erosion scarp

Fig 168. The formation of consequent and obsequent erosion fault scarps

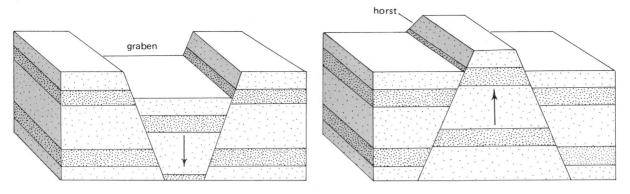

Fig 169.   Block diagrams to show the structure of horst and graben landforms

Fig 171.   Uganda: (Right) aerial photograph of the eastern fault-scarp of the western arm of the Great Rift Valley at Lake Mobutu. The rivers have hardly begun to erode the edge of the scarp. Note the deltas extending into the lake.

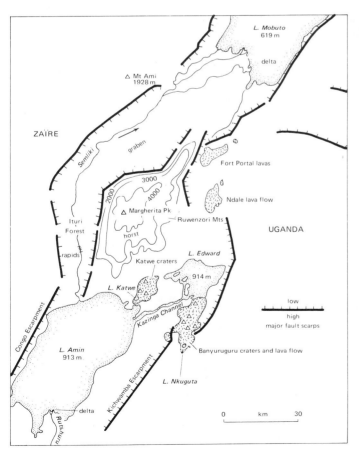

Fig 170.   Landforms due to faulting and the drainage system of the Ruwenzori region, Uganda—Zaïre border

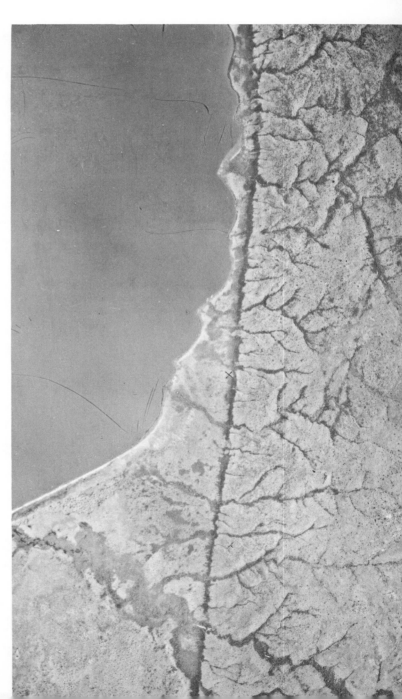

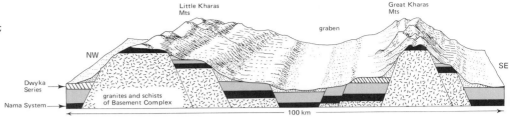

Fig 172. Block diagrams to show the horsts of the Little and Great Kharas Mountains and the graben between them; Kalahari Desert, Namibia

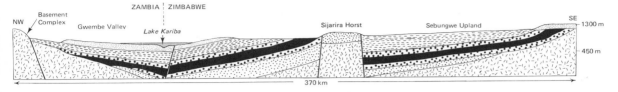

Fig 173. Cross-section of the Gwembe Valley and Sebungwe Upland showing the effect of faulting. The black band represents the Wankie shales and coal measures

The Great Rift Valley

The Great Rift Valley of Africa provides some of the most spectacular examples of faulted landscapes in the world. The valley extends some 7200 km from the Jordan section to the coastal region of Mozambique near Beira, with 5600 km lying within Africa. From the Jordan trough, whose floor in the Dead Sea lies 395 m below sea-level, the valley is marked by the fault-bounded Gulfs of Aqaba and Suez. Between lies a highland mass of granitic and metamorphic rocks which fall in a steep fault-scarp to the Gulf of Aqaba and in a series of step faults to the El Qa'a coastal plain in the west. The Red Sea is also a huge submerged rift valley with long straight shorelines and narrow coastal plains. Its width, 355 km at the maximum, is due to the gradual separation of the Arabian tectonic plate from Africa. Here the edge of the Nubian Desert in Sudan is marked by a dissected fault-line escarpment 600 m–1200 m high which overlooks a coastal plain 15 km–35 km wide.

In northern Ethiopia the valley is seen in the wide fault-bound triangular Danikil Plains. Here the western scarp slopes have been dissected into numerous foothills which rise steeply to uplifted plateaux capped with basalt lava 3000 m above the lowland plain. The eastern scarp faces are generally lower, the Ahmar Mountains rising about 2000 m above the plain, and decrease in altitude eastwards. Two interesting tectonic features occur on the Danikil lowlands: the Kobar Sink, a fault-bounded graben measuring 170 km by 65 km, its bed being 116 m below sea-level, and the Afar Alps, a long dissected horst lying parallel to the shores of the Red Sea.

Southwards the Great Rift Valley narrows to an average

Fig 174. Kenya, Lake Magadi: this alkaline lake is situated in the southern section of the Great Rift Valley and lies between low fault scarps within the Rift. The mountains are of volcanic origin.

width of 50 km and divides the much faulted western and
eastern highlands of Ethiopia. The valley is here marked by a
series of small lakes which form centres of internal drainage—
Lakes Zwai, Langana, Abaya, Shamo and the ephemeral L.
Stefanie. The Omo River trough is marked by low faulted hills
and leads into the Kenya section of the Rift Valley.

The Central Rift of Kenya (also called the Eastern or Gregory
Rift) is marked near the northern boundary by a series of low
faults, L. Turkana occupying a downwarped rather than a strongly
faulted depression. It is in the Central Highlands of Kenya where
the Great Rift Valley displays its most impressive features.
Here the valley floor is some 60 km wide and 400 km long and is
bounded by sharply defined fault-scarps varying in height from
500 m to 1000 m above the floor. The bounding faults occur in
series and are normal faults between 20 km and 30 km long with
throws sometimes exceeding 1500 m. Both step faulting and
fault splinters are common. The faults in this section are considered younger than those of the western arm and are mainly
of Pleistocene age, faulting having occurred at various times.
The valley is at its most impressive near Nakuru where the
Aberdare Range to the east towers 2200 m above the valley
floor (Settima 3999 m, Kinangop 3906 m) and is faced in the
west by the Mau escarpment (1200 m above the floor). The
great heights of the fault scarps here are due to their being formed
on the slopes of the huge Kenya dome (p. 80). Further south
near L. Magadi they are much lower. Fault grids appear on the
floor of the valley north of L. Magadi and individual blocks
have been formed in the Sike's Grid near Mt Suswa. In the
Londiani region the western wall of the valley is broken by
transverse faulting which leads to the Kavirondo Gulf of
L. Victoria (see Map 13, p. 141).

The minor faulting which has occurred on the floor of the
Rift Valley has produced low depressions which have been
filled with small shallow lakes, none deeper than 16 m, such as
Elementaita, Nakuru and Naivasha and the intensely saline
lakes Magadi and Natron (p. 107). Strong faulting occurs in the
deep trough occupied by L. Eyasi with some spectacular
scarps to the south of the lake. In general, however, the fault
scarps are lower in Tanzania and are often replaced by trapdoor faults—low scarps caused by a gentle dipping or sagging
of the surface rocks (Fig 163).

In central Tanzania the Rift Valley is replaced by a complicated horst and graben landscape described above. North of L.
Malaŵi, however, the rift structure appears again and here links
with the Western Rift Valley and the Malaŵi rift zone. The
southern section of the western arm has several major faults
running almost parallel with each other in the horst and graben
landscape of the Ufipa Highlands and the L. Rukwa depression.
Faulting has produced submerged ridges in L. Tanganyika and
the lake is bordered by low fault scarps some 150 m above
water level (p. 70). The country to the north around L. Kivu
and the Ruzizi Range is rugged and broken and partly drowned
by the lake to form rias (p. 121). L. Amin, in contrast, occupies
a well-marked rift valley which divides northwards into the
L. George lowland and the Semliki trough. L. Mobutu lies in a
trough bounded by well-marked fault-scarps which gradually
decline in height until they are replaced by the zig-zag fault
zones of the Nile region.

In Malaŵi the Rift Valley forms a great trough some 80 km
wide and 650 km long. The faults bounding L. Malaŵi also
display a zig-zag pattern which accounts for the irregular shape
of the lake (p. 70). To the south lies the Shiré rifted region with
two main series of fractures, one trending south–southwest
between the southern lake shore and the Murchison Rapids,
and the second aligned south–southeast between the rapids and
Nsanje. The flat-topped plateaux here are not horsts but ancient
blocks of plutonic and metamorphosed crystalline rocks which
have been uplifted and bevelled (Fig 67). The western edge of

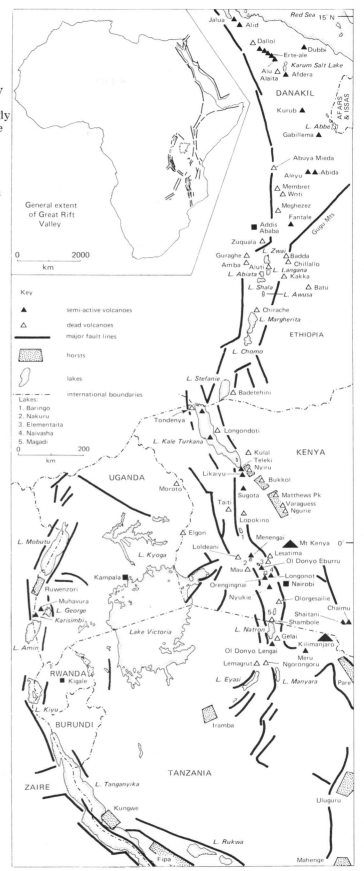

Fig 175.　Major landforms associated with the Great Rift Valley of
East Africa

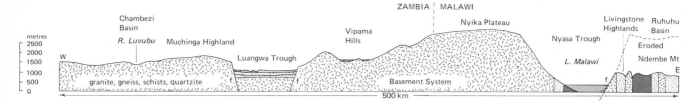

Fig 176. Cross-section through northern Malawi east-west into
Zambia showing the effect of faulting

the Zomba Plateau is bounded by one of the main Rift Valley
faults.

Between the Zambesi and the Pungwe rivers in *Mozambique*
the valley continues in the Urema Trough, a faulted marshy
depression up to 65 km wide drained by the Urema River.

The Rift Valley also has a western arm in *Central Africa*
which is marked by the Luangwa trough in Zambia and the
Zambesi valley as far as the Kariba Gorge (Fig 173).

Scientists have produced several theories to explain the
formation of the Great Rift Valley of Africa. The tensional
theory in which a wedge-shaped block sinks slowly between
two normal bounding faults as lateral pressure is released was
applied by J. W. Gregory to the Kenya rift section. This theory
would fit the conditions produced by crustal plates drifting
apart from each other. B. Willis and E. J. Wayland favoured the
compression theory in which compression deep in the earth's
crust caused crustal blocks to ride up a fault plane to a high
angle (Fig 165b); the leading edge cannot be supported and
collapses to bury the fault line so that the resulting slopes
appear at the surface as normal fault slopes. This idea, however,
would not fit in with the theory of plate tectonics.

Another idea, suggested by F. Dixey, was that the rifts were
related to the basin and swell structure of Africa. The Rift
Valley in Malawi is, according to Dixey, an extension of the
Mozambique Channel, a geosyncline lying between the upwarped swells of East Africa and Madagascar. As the mainland
rose, ancient fractures formed zones of weakness which led to
rifting. Again, this explanation does not fit in with the theory
of plate tectonics.

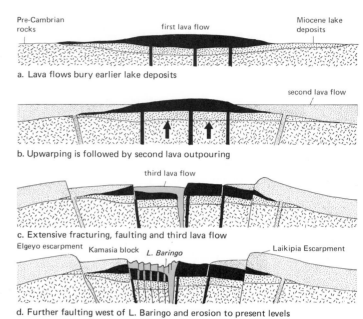

a. Lava flows bury earlier lake deposits

b. Upwarping is followed by second lava outpouring

c. Extensive fracturing, faulting and third lava flow

d. Further faulting west of L. Baringo and erosion to present levels

Fig 178. Four diagrams to show the successive stages in the formation
of the Rift Valley in the Lake Hannington area of Kenya. Distance
across the sections is about 160 km

Fig 177. Tanzania: view from the slopes of Ol Denyo Lengai, an
active volcano, showing part of the steep fault scarp of the Rift Valley.
Note the deep gulleys caused by occasional rainstorms in this usually
semi-arid region. In the background stands Mount Gelai (2 947m) which
overlooks Lake Natron.

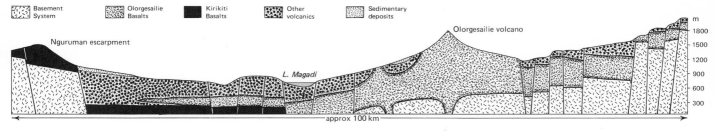

Fig 179. Cross-section of the Great Rift Valley along latitude 1°45'S. normal fault at left
Note the step faults at right and the considerable downthrow in the

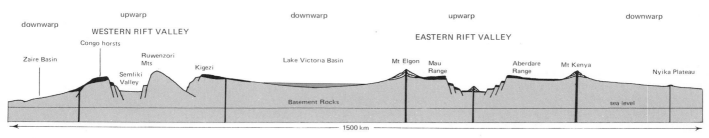

Fig 180. Generalised section across East Africa showing the effects of warping and subsequent faulting

The theories of these geomorphologists can be applied in part to some sections of the Rift Valley for there is evidence of both tension and compressional faulting. In East Africa at least, the Rift Valley has formed on a huge upwarped swell or tectonic arch which formed during and after the Tertiary period. The central portion of this arch sagged to form the downwarped region occupied by Lake Victoria (Fig 180) while the flanks have been upwarped and faulted to create the western and eastern arms of the Rift Valley.

Recently, the work of H. Hess and others on sea-floor spreading and plate tectonics has indicated a definite link between these forces and the formation of the Rift Valley. The most recent geophysical research indicates that there were three phases of rifting: during the Triassic (200 million years ago), during the Cretaceous (80–130 million years ago), and during the late Cainozoic (over the last 20 million years) when the eastern Rift Valley was fully formed. The major period of faulting began only some 11 million years ago and this was followed by lesser faulting on the valley floor during the last 3 million years. A further period of uplift accompanied by faulting occurred within the last million years. This last phase of faulting has been linked by geologists to the break-up of Gondwanaland and the tension set up in the earth's crust by the subsurface movement along plate lines (p. 10).

Questions

1. Describe and explain the mode of formation of the following tectonic features referring to examples in Africa:

   a fault-scarp; a fault-line scarp; a thrust fall; step faults; a horst; a graben.

2. Write an essay entitled, 'The Great Rift Valley of Africa: Some Theories Regarding its Origin'.

3. With the aid of sketch-maps and diagrams, describe the extent of the Great Rift Valley of Africa and discuss the theories regarding its origin. You should relate the formation to the theory of plate tectonics.

4. Discuss the landforms associated with rift valleys in Africa.

13   Volcanic Landscapes

Although Africa has been subject to crustal warping and severe faulting in the Rift Valley region, the continent is a generally stable one which has existed as a rigid block since pre-Cambrian times. In several regions, however, volcanic activity has disturbed the long quiet processes of landscape evolution to produce magnificent peaks and extensive lava cappings.

Volcanic Zones in Africa

The major areas of vulcanicity are chiefly associated with the crustal fractures of the Rift Valley and the fault zones of adjacent regions in central Kenya, northern and southern Tanzania, northeastern Zaïre, the adjoining parts of Uganda, and Ethiopia. The first zone includes Africa's highest peak, Kilimanjaro (5894 m) and also Mounts Kenya (5199 m), Elgon (4322 m), and Meru (4566 m). In addition thick cappings of lava cover many thousands of square kilometres of the planed surfaces. The Zaïre–Rwanda–Uganda region includes the Mufumbiro Mountains with the extinct Muhavura (4113 m), Gahinga and Sabinio, and further south near L. Kivu, Nyamlagira and Niragongo, the only active volcanoes in the region (Fig 129). Other dead volcanic cones include Karisimbi, Visoke and Mikeno. The Gwasi region south of the Kavirondo Gulf in Kenya has experienced several eruptions associated with local faulting to produce the Rangwa volcanic complex and the Kisingiri volcano (Fig 186). In Ethiopia, extensive areas are covered with great thicknesses of plateau basalt released through crustal weaknesses caused by rift faulting and there are numerous small cones, dykes, sills and hot springs along the Danikil faults.

In West Africa igneous activity is associated with higher relief on the Jos Plateau, the Bamenda and Cameroon highlands, the Opon-Mansi Mountains of Ghana, active Mt Cameroon (4070 m), and on the islands of Fernando Po, Principe, São Tomé and Annobon. Basic intrusive lavas have formed the Sierra Leone Mountains (a huge lopolith) and Liberia's Cape Mount and there are numerous dyke forms in the Futa Djallon of Guinea and the Tarkwa region of Ghana. Volcanic massifs are located in the Aïr, Tibesti and Hoggar.

The volcanic activity in East and West Africa occurred largely during the last 3 million years during the Tertiary and Quaternary periods. In southern Africa volcanic activity occurred much earlier, some 250 million years ago during the Carboniferous Period. This vulcanicity did not produce impressive mountain peaks but landscapes whose surface is modified by intrusive forms and lava flows. In the Transvaal, lava intruded into overlying sedimentary rocks, broke to the surface via pipes and spread for nearly 500 km over the plateau. The Drakensberg region of Lesotho and Natal experienced extensive lava flows which covered the Cape Sandstone beds to a depth of 1500 m; lavas of the same age are also widely exposed in Mozambique, Zimbabwe, Botswana and Namibia.

Volcanic massifs also occur in the Comores Islands with peaks rising to 2475 m on Grande Comore and also on Madagascar — the massifs of Montagne d'Ambre (1360 m) and Tsaratanana (2880 m) in the north, and the extensive Ankaratra region (2644 m) of the centre which resulted from severe faulting and magma release.

Volcanic Activity

The formation and location of volcanic landforms depends on the supply of magma from subterranean reservoirs and the position of zones of weakness in the crust. At a depth of 100 km beneath the surface, temperatures reach 1500°C, enough to melt any rock at the surface but at such a depth the pressure of the overlying crust keeps the rock in a plastic, almost solid state. Once this pressure is relaxed, due to crustal weakness such as folding or faulting, the superheated rock or magma begins to move upwards. Accompanied by liquids and gases it runs along the fissures, expanding and becoming lighter. If the magma reaches the surface it emerges as lava with a temperature of between 1000°C and 1200°C.

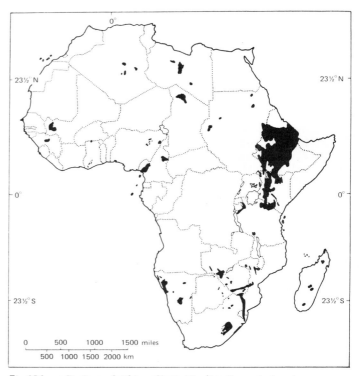

Fig 181.   Regions of Africa affected by lava flows during the Cretaceous (largely south of the equator), Tertiary and Quaternary periods

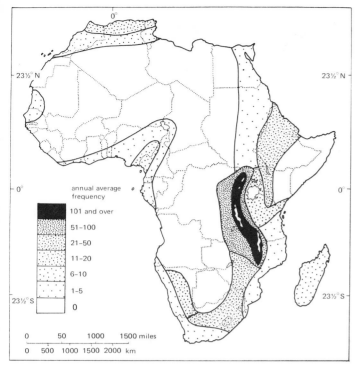

Fig 182. The annual average frequency of earthquakes in Africa. Note the very high frequency associated with the Rift Valley faulted zone in eastern Africa

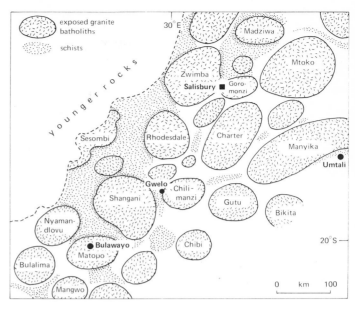

Fig 184. Granite batholiths exposed at the surface in Zimbabwe. Their surfaces have been eroded to form wide areas of level pediplains with domed inselbergs and castle koppies

Magmas and lavas are of two types. Basic types are very fluid, poor in silica but rich in magnesium and iron; they rise rapidly to the surface and normally pour out or extrude without much disturbance. Acid types are rich in silica, solidify at high temperatures and move slowly through the earth's crust. Accompanying gases and steam escape easily, pressures build up, and violent surface explosions often occur.

Volcanic eruptions are merely the last stages in the processes called vulcanism which begins with the formation of plutonic magma and ends with its solidification beneath or on the surface of the earth. Volcanic landforms are strongly influenced by the

nature of the magma or lava and fall into two groups: intrusive forms and extrusive forms.

Intrusive Igneous Forms

As well as penetrating zones of weakness, magma may melt the surrounding rocks and absorb huge blocks of rock, a process known as stoping. Such blocks, called xenoliths, are sometimes seen in the solidified magma. Magma may also force apart rock beds by pressure.

The largest intrusive forms are batholiths formed from magma which solidified at great depths. Batholiths may be tens or hundreds of kilometres across and often form the exposed cores of mountain ranges. The magma cooled slowly often forming large-grained granite (p. 19) and metamorphosing

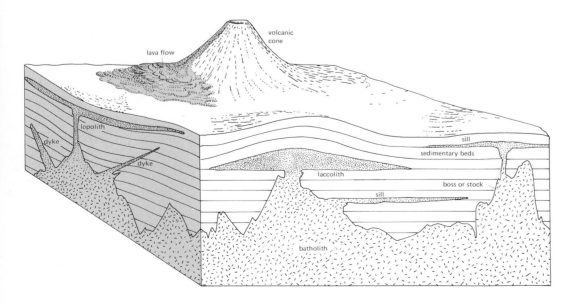

Fig 183. Block diagram to show the major intrusive and extrusive features due to volcanic activity

Fig 185.   Kenya, the Thika Falls: caused by a resistant volcanic dyke

f. Further intrusion (6) forms dykes and surface lava flow. Erosion to present levels.

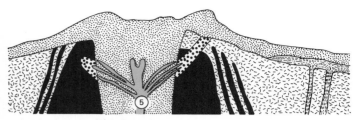

e. Another intrusion (5) in form of cone dykes

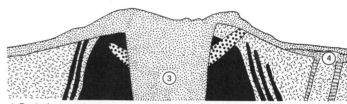

d. Explosive eruption occurs removing much of original rocks (3). Breccia-type lava pours out onto surface. Later intrusion (4) forms dykes and sills.

c. Erosion of up-domed Pre-Cambrian surface

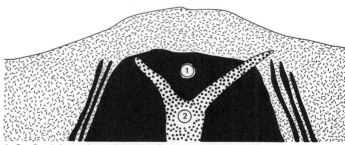

b. Pre-Cambrian rocks up-domed by first magma intrusion (1) which is followed by a second intrusion (2).

a. Original surface of Pre-Cambrian rocks

Fig 186.   The evolution of a volcanic landscape: the Rangwa volcanic complex southwest of Kisumu and the Kavirondo Gulf, Nyanza Province, western Kenya

surrounding rocks in its aureole into schists and gneisses. Columns of magma called bosses or stocks sometimes extend upwards from the batholith into weak sections of the crust (Fig 183). Such deep-seated forms are only exposed at the surface after long periods of denudation.

Considerable areas of batholithic granite are exposed in the Transvaal, Natal and Namaqualand. Over most of central Zimbabwe granitic batholiths surrounded by metamorphosed elliptical zones of schists are common (Fig 184) and the Old Granite of these batholiths form much of the basement series of southern Africa. The intrusive domes of Tamgak and Baguezane in the Aïr Mountains rise to 1800 m and these batholiths have been exposed to erosion since the Tertiary period. Extremely old granites of the Basement System are also intruded into the pre-Cambrian rocks of the plateaux of East Africa. Central and northern Tanzania are underlain by a granitic shield which is occasionally exposed at the surface. In Sukumaland in Tanzania the surface is underlain by a batholith with bosses exposed at the surface as tors and castle koppies (p. 29). In central Tanzania and northern Uganda basement granites stand up as inselbergs and ranges of mountains, e.g. the Labwar Hills and the Parabong Range in eastern Acholi which rise to 6000 m above sea-level.

Magma intruded between bedding plains may develop a saucer-like form called a lopolith. This is due to the tremendous weight of the magma and the weakening of the rock strata beneath by stoping. The Bushveld Igneous Complex of the Transvaal which was intruded nearly 2000 million years ago is a huge lopolith formed by several igneous intrusions which became so heavy that they caused the underlying rock to sink. In some places the lava reaches depths of 10 km and the maximum east–west length is 480 km (Fig 188).

In the Great Dyke of Zimbabwe (Fig 187) the intrusives have formed parallel but discontinuous strips dipping inwards towards the central spine in a lopolith feature. The 'dyke' extends for some 560 km north–south through the centre of Zimbabwe and averages 11 km in width. It forms a distinctive series of ridges rising in the Umvukwes Range north of Salisbury to nearly 500 m above local levels. Dykes and sills are common in Zimbabwe and are the result of extensive volcanic activity which occurred at intervals between 150 and 2000 million years ago.

96

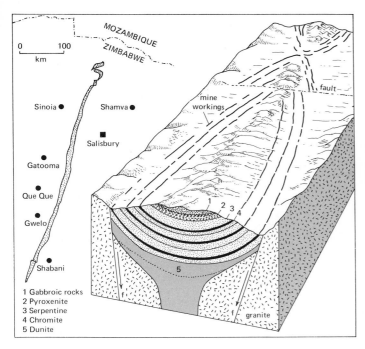

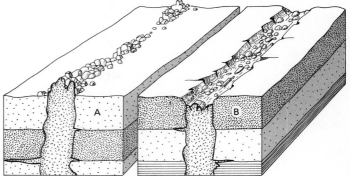

Fig 189. Block diagrams illustrating the effects of a volcanic dyke on surface landforms. *Left:* The rock at A is less resistant to erosion than the solidified lava of the dyke and a broken ridge results. *Right:* The rock at B is more resistant than the dyke and the latter is eroded to form a long depression in the surface.

Fig 187. Sketch-map and block diagram of the Great Dyke of Zimbabwe. This is really a lopolith with plutonic rocks about the same age as those in the Bushveld Igneous Complex (Fig 188). It is 515 km long and from 5 to 10 km wide. The first stage of the Great Dyke's formation was the intrusion of magma in lopolith form in a series of injections one after the other. Before each injection of magma solidified the heavier minerals sank to the base of the injection leaving mineral bands, e.g. the chromite. The weight of the magma caused faulting and sinking in a graben. Erosion then reduced the lopolith to the present dyke-like form and exposed the mineral bands on the flanks of the central ridge

A similar feature to the lopolith is the laccolith which forms when viscous (slow-flowing) lava builds up along a bedding plane. If the magma cannot break through to the surface the overlying beds arch up to form a dome. Some of the impressive mountain features of southern Malaŵi are due to this doming process. Intrusions of lava which now form the massifs of Zomba, Mlanje, Machema, Nasolo and Michesa were once buried beneath thick sediments now stripped away. The laccolith form also exists below the base of Mt Cameroon.

Sills form where the magma penetrates in sheets along the bedding planes of sedimentary rocks by partial melting of the beds or by forcing them apart. Sills lie almost horizontally, are of varying thickness and may extend for hundreds of kilometres. Dykes on the other hand are vertical or near vertical intrusions of magma which cut across sedimentary rock beds, usually penetrating fault or fissure lines to form walls of solidified magma. They may split or join or radiate from the base of volcanoes if the outlet is blocked. Many such dykes are called a swarm. If exposed at the surface, a dyke may form a wall or trench depending on the comparative resistance of the surrounding surface rock (Fig 189). Such differential erosion is seen in Lesotho: where dykes occur in harder sandstones they have been eroded into trenches but where they transgress softer shales they stand up like low broken walls. In many parts of the Karoo, dykes 6 m–10 m wide often form long lines of small hills or koppies and in the Isizwa Mountains of Griqualand East bands of dolerite form a sill many hundreds of metres thick. In the upper Niger basin the generally unbroken plain is interrupted by a low line of hills less than 185 m high formed by the Kouroussa Dyke which extends for 160 km in a SSE–NNW direction.

Rings of dykes forming round a central plug are common in Malaŵi. The central core of the Zomba Plateau is surrounded by circular dykes and the Chambe massif to the northwest of Mlanje is another ring dyke complex. The Rangwa complex of Kenya (Fig 186) has a similar feature and Tororo Rock in eastern Uganda is an igneous intrusion with a 'collar' of dykes.

Dykes and sills are a common intrusive feature of the sandstones of the Futa Djallon highlands in Guinea where they form numerous ridges and rocky ledges over which flow impressive

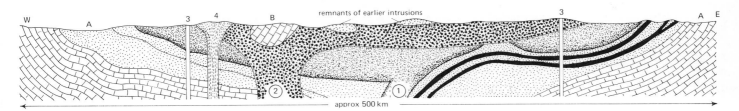

Fig 188. Diagrammatic section across the Bushveld Igneous Complex, southern Transvaal, South Africa. This huge lopolith was formed in the following manner: An intrusion of basic magma (1) called norite penetrated the sedimentary rocks of limestone and sandstone (A). A second intrusion of acid magma, the Red Granite (2), penetrated the first intrusion. The weight of these magmas caused the sedimentary rocks to sink. Erosion exposed the acid magma over a wide area in the centre of the complex. Xenoliths of sedimentary rock (B) were trapped in the magma. After this plutonic phase, a volcanic phase began, with narrow volcanic pipes (3), and lavas and syenites being extruded (4). See Fig 52 for the extent of this igneous complex

Fig 190. Zimbabwe, Inyangani Mountain (2 593m): the highest point in Zimbabwe; it is a thick dolerite sill overlying granite

Extrusive Igneous Forms

When the magma reaches the ground surface it emerges as a lava flow usually accompanied by the release of a variety of gases including steam in large quantities which creates tremendous explosive pressures, plus carbon dioxide, sulphur dioxide, ammonium chloride, hydrogen sulphide, and hydrochloric acid. Deposits of sulphur crystals often surround small gas vents. Some of the gas may be trapped in the lava forming small gas bubbles and the result, called pumice, looks like a hard rock sponge. Large quantities of pumice stone occur in the volcanic areas of the Rift Valley of Kenya and Tanzania. The holes may have chemicals precipitated into them which become fossilised into rainbow-coloured agate and calcite minerals to form amygdales, a common feature of the basaltic lavas of the Drakensberg and the Victoria Falls region.

Pieces of rock ejected during an eruption are termed ejectementa or pyroclasts. They consist of blocks of surface rock, solidified lava pieces, cinders (*scoriae*), pumice stone, and solid volcanic bombs. Such pyroclasts often form the unstable porous slopes of the upper parts of volcanoes.

The surface features formed by eruptions depend on the nature of the raw material ejected and the type of eruption. Basic lavas such as basalt which contain up to 60 per cent silica are very fluid and form low features whereas acid lavas with a high silica content of up to 70 per cent build up tall volcanic cones. As a basic lava flows along the ground its outer surface is cooled and a skin forms, while beneath this the lava continues to flow causing the surface skin to form wrinkles or rope-like folds. Acid lavas such as rhyolite and obsidian are extremely viscous, move very slowly and are soon cooled into a blocky structure. The lava flows from the active Teleki volcano in Kenya are of the blocky type. Rope-like lavas are a particular feature of the Nyamlagira volcano in southwestern Uganda where the Saki Bay lava flowed for some 24 km in a 5 km-wide tongue in 1942, moving an estimated 1·5 billion tonnes of molten rock.

waterfalls. Faulting in the Tarkwa region of Ghana has also allowed penetration of laccoliths, sills and dykes over a wide area and the cores of many hills in the Opon-Mansi Hills are formed of intrusives. The highest peak in Sierra Leone, Mt Bintimani (1948 m) is composed partly of sill intrusions. Inyangani Mountain (Fig 190), the highest part of Zimbabwe, is formed of granite capped by a thick dolerite sill and in Kenya sills and dykes are common in the Thika district where they cause numerous rapids and waterfalls, e.g. Thika Falls (Fig 185).

Fig 191. Nigeria: a granite dyke on the eastern edge of the Jos Plateau, southeast of Jos. An upper tributary of the Gongola River forms a waterfall as it leaves its V-shaped valley on the surface of the dyke. The braided channels and ill-assorted deposits in the foreground result from the abrupt change of slope

Fig 192. Kenya: the Aberdare Range forms the eastern flank of the eastern Great Rift Valley and lies north of Nairobi. The summit in the background is nearly 4 000m above sea-level, beyond it there is a drop of 2 200m to the floor of the Rift Valley. This is the surface of a thick layer of lava which caps the Aberdares

Lava Flows

When basic magmas well up along a fissure and a continuous supply is assured from a batholith, then the whole landscape may be completely buried under an enormous sheet of lava and later extrusions may greatly increase the depth of the original outflow. After the new surface has cooled and hardened to a level plain the forces of erosion will begin to create a new landscape. Extensive lava flows are a common feature in Kenya occurring over vast areas in the L. Turkana region and on the Laikipia and Athi plains southeast of Nairobi. The Loita Plains of the Narok region of Kenya are also an unusually level flat surface composed of Tertiary ashes and lavas, while thin lava sheets occur in the Kitale area. In the Nakuru–Lake Hannington area the lava piles are 2100 m thick, while on the Aberdare and Mau ranges the lava cappings add a further 1000 m to the height.

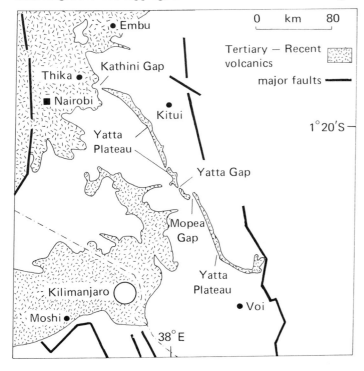

Fig 193. The extent of the Yatta Plateau and Tertiary to Recent volcanic lavas in central Kenya and northern Tanzania

The Yatta Plateau of Ukambani, east of Nairobi, is a natural wall of lava some 90 m high and from 3 km to 15 km wide, formed when lava seeped out of a 240 km-long fissure. In Uganda lava flows blocked the drainage flow to create L. Amin (Edward) (p. 70).

During the Tertiary period basaltic lava flowed to cover the Biu plateau of Nigeria and to fill the valleys eroded in the granite of the Jos Plateau. When the lava had solidified in the valleys it was more resistant to erosion than the granite between. The granite was eroded more quickly leaving the former valley basalts as resistant hills, a good example of inversion relief (Fig 195). The weathered basalt now forms deep clay regoliths. Later flows during the Cretaceous about 50 000 years ago also buried the southeastern parts of the Jos Plateau. The basalts of the Biu Plateau, on the Jos surface, and in Cameroon and on the Cape Verde peninsula are all of about the same age.

In the Drakensberg Range of Lesotho and Natal, Jurassic basalt lavas, in places 1500 m thick, now cap the sandstone beds. As 150 million years of erosion have taken place since the eruptions, the lava must have been considerably thicker than at present. Here there are lava beds from 1 m to 50 m thick laid down almost horizontally on top of one another. The upper lavas welled up from fissures but the earliest beds were caused by explosive eruptions and are filled with fragmented rocks or breccias. In the eastern Transvaal and the Lebombo Mountains of Swaziland and Zululand similar basalts are overlain by later thinner sheets of more acid lavas, the whole layer being 1000 m thick. The Kaokoveld area of Namibia has similar basaltic lava sheets.

Volcanoes

Volcanoes are unique since they are not caused by the normal processes of erosion, transport and deposition like other landforms but are due to constructive forces of relatively short duration and are altered and removed by either violent explosion or by gradual erosion. The nature of the volcanic raw material and the type of eruption—explosive or gentle—are the chief controls of volcano forms.

Thus, when fluid lavas erupt through a narrow vent instead of a long fissure a low dome shape is formed with low slope angles instead of level cappings. A typical example of this form, termed a shield volcano, is extinct Nyamlagira in northeastern Zaïre.

Occasionally the surface rocks may be ruptured by gas

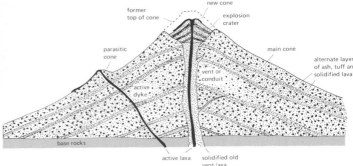

volcanoes include Mounts Cameroon, Kenya, Meru and Kilimanjaro. Kilimanjaro is a dormant volcano its base covering 2800 km² which has evolved from three volcanic mountains (Fig 201). The oldest rises to Klute Peak (4000 m) on the Shire Plateau, a shallow caldera (see below) on the western side; the second centre of activity is marked by frost-shattered Mawenzi (5150 m); between these two lies the most recent summit of Kibo (Uhuru Peak 5895 m) which is also a caldera. Inside the Kibo caldera is a nested cone and a crater with several sulphurous fumaroles or ejection pits.

Fig 194.   Zaire: a blocky lava flow emitted by the Virunga volcanoes

pressures which create a small crater or pit surrounded by low ridges of pyroclasts. These low profile features called explosion vents, ash rings, explosion pits, gofs or *maar* (Ger.) may sometimes be occupied by shallow lakes. In southwestern Uganda in the Fort Portal region of the Ruwenzori National Park there are over 200 explosion pits, some filled with lakes such as Lakes Kitinda and Mirambi, others dry or marshy such as Rutunguru, Itunga and Kyangabi craters. Here, Lake Katwe is an important source of local salt from the evaporated crater lake. Similar explosion craters have been blasted through the Nubian Sandstones in the Darfur region of the Sudan.

The violent ejection of pyroclasts and ash from a vent may continue for weeks or even months. Lava fountains into the air in liquid form and returns to ground-level as semi-solid or solid cooled pieces called lapilli if very small, or scoriae (cinders). The fine ash and cinders descend in great clouds and an ash cone is built up whose angle of slope varies between 25° and 35° for fine ash to about 40° for cinders. Such a cone may reach a great height after only a few days of activity. The cone of Mt Elgon on the Kenya-Uganda border is largely an ash and agglomerate cone and the Chuyulu Hills in Kenya are formed by a line of ash cones. Ash cones have also formed within the crater of Mt Meru in Tanzania and as a parasitic cone on the northeastern flank of Mt Longonot in the Kenya Rift Valley. The Jos Plateau in Nigeria has several small ash cones on its surface.

Most volcanoes, however, consist of alternating beds of ash and lava which build up into a cone shape with a central vent. These are called strato- or composite volcanoes (Fig 196). Some layers are of volcanic ash mixed with scoriae and fractured rock called volcanic breccia or agglomerate while the contents of other layers may be derived from surrounding rock from the early stages of eruption. Composite or strato-

Fig 196.   Diagrammatic section of a composite or strato volcano

The craters of such volcanoes may be quite extensive—Mt Elgon's is 8 km across—and sometimes contain lakes. In West Africa L. Bosumtwi is a circular crater lake 32 km southeast of Kumasi and a similar feature is the Panyam crater lake in the Bauchi Province of Nigeria. Some craters are almost perfectly circular in shape and there are several examples of this in the active and dormant volcanoes which lie along the fault lines of the Danikil depression in Ethiopia. Erta-ale at the northern end of the Danikil Desert is a circular, dish-like volcano which erupted in 1960. This was an effusive eruption, the lava welling out quietly like treacle or hot tar and filling the shallow crater to form a rare lava lake.

The craters of volcanoes are sometimes enlarged to form calderas. Calderas may be caused by a pressure explosion which shatters the top of the volcano. Within this new crater nested cones may form only to be destroyed by later explosions resulting in a complicated volcanic region of low profile. Mt Meru in Tanzania has had the apex of its cone blown away with later cones developing inside the new basin. Other calderas are caused by subsidence of the rim of the original crater followed by inward collapse. Large sections of the rim fall into the magma cauldron and are remelted thus enlarging the crater. The largest example of a caldera in Africa is the Ngorongoro Crater in northern Tanzania which measures 22 km across and contains a small alkaline lake. The surrounding region is called the Crater Highlands because of the groups of calderas and volcanic craters covering 1800 km². Other examples of calderas include Menengai near Nakuru in the Rift Valley and Napak in southern Karamoja (Fig 198).

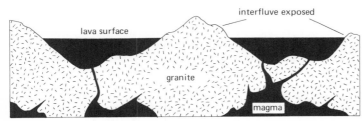

Fig 195.   The inversion of relief caused by vulcanicity, e.g. on the Jos Plateau, Nigeria. *Left:* Lava extrudes to bury the valleys of the granite but leaves the interfluves exposed. *Right:* The granite is less resistant than the solidified lava; the latter now forms the highland, the granite forms the valleys.

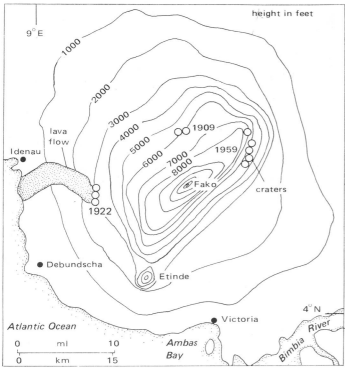

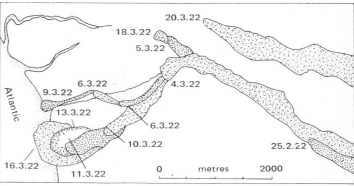

Fig 197. Mount Cameroon, West Africa. *Left:* Contour sketch-map showing craters and lava flows. *Right:* The progress of the 1922 lava flows. Using the scale, estimate the speed of the lava movement. Mount Cameroon is a typical strato or stratified volcano (Fig 196) with a base measuring 48 x 33 km. The base of the volcano dates from the Upper Cretaceous period. Etinde Peak was formed in the late Tertiary when there were large emissions of basaltic lavas. Later activity occurred in the Miocene and Pleistocene periods. During the nineteenth century Mount Cameroon erupted five times and again in 1909, 1922, 1954 and 1959

Fig 198. (Left) The evolution of the Napak caldera, southern Karamoja, eastern Uganda

Fig 199. Tanzania: Ol Doinyo Lengai photographed during the eruption of 1966. The alkali-rich lava is black when it first flows but is converted to white sodium carbonate on contact with moisture in the air. The symmetrical cone is formed by explosive eruptions

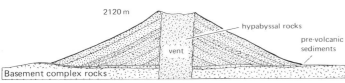

a. Formation of the Lower Miocene volcano

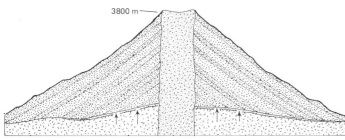

b. Napak attains maximum height during Lower Miocene. Basement rocks updomed

c. Major faulting of domed basement rocks causes collapse of Napak volcano

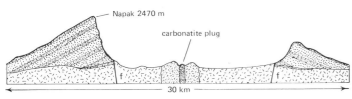

d. Eroded remnants of Napak caldera

Fig 200. Uganda, Tororo Rock, Bukedi Province: this is an eroded plug of a volcano which rises abruptly above the surrounding plains

Volcanoes may also be destroyed by the more gradual processes of erosion. Cones composed of ash or agglomerate are soon rilled and gulleyed by rainwater but are not always easily removed since their porous slopes allow water to sink through the surface to form a complicated underground drainage system. When the volcano has been reduced considerably by erosion the resistant solidified lava of the throat and vent are exposed to leave a rugged volcanic neck or plug. Mt Kenya (5199 m) at one time exceeded 6200 m in height but erosion has produced the ice-shattered resistant pinnacle of today (Fig 144). Exposed volcanic cores also form an irregular mountainous area between the Great Rift Valley and the Kano Plains in Kenya with peaks at Tinderet (2641 m), Timboroa (2894 m) and Loldiani (3031 m).

Volcanic Hot Springs

Hot springs are frequently associated with the volcanic regions of Africa. Water from rain, snow or plutonic sources seeps through cracks or along ash and agglomerate layers and occasionally collects in underground caverns. During its passage it may come into contact with heated rocks and later emerge as a spring of warm or hot water. The waters carry chemicals in solution which are often deposited in thick layers along the lower parts of the subterranean water course or around the rim of the spring exit. The saline Maji Moto (Swahili—hot water) which flows into L. Manyara in Tanzania issues at a temperature of 74°C. A similar hot spring, also called Maji Moto, occurs near Nakuru and there is a large carbonate spring near Eldoret. Not all hot springs contain chemicals in solution, however, and the Ikogosi hot spring in western Nigeria which issues at a temperature of 38°C is fresh water.

Sometimes the percolating water is transformed into superheated steam which, mixed with other gases, emerges as a steam jet through small vents termed fumaroles. Numerous steam jets occur along parallel fissures in the Rift Valley between lakes Elementaita and Naivasha while there is a large steam jet within the crater of Mt Longonot. Where fumaroles emit strongly sulphurous gases they are called solfataras and many of the fumaroles of the Hoggar and Tibesti are of this kind. Along the fault lines bordering the Danikil lowlands of northern Ethiopia, chemically coloured hot springs are caused by water seeping into faults and cracks, contacting molten rock, and slowly percolating through minerals and salts. As the hot water cools its minerals begin to crystallise to form minor landforms like the travertine of limestone karstic regions.

In areas where there has been recent alluvial deposition

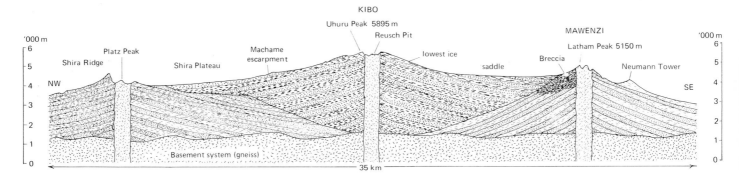

Fig 201. Diagrammatic section across the Kilimanjaro volcanic region

Fig 202. Kenya, Mount Longonot: a caldera in the central Rift Valley.-
Note the sharp rim of the caldera which extends for about 15 km, the
thick lava flows in the foreground, the symmetrical ash cones on the
right, and the deep scoring of the caldera sides by gulleying

explosive eruptions by steam and other gases sometimes produce low cone-shaped clay mounds about 2 m high and about
20 m in diameter called mud volcanoes, e.g. in the Sekenke area
of Tanzania. These are not major volcanic features, however,
and are ephemeral, being soon washed away.

Fieldwork

Visits to areas of volcanic activity are possible for students in
the vicinity of the Rift Valley, the Aberdares, Mt Kenya, Elgon,
the Jos Plateau, etc. Such itineries should be well-planned to
cover all aspects of vulcanism. The following are important
activities: landscape and landform sketching, collection of
samples of scoriae, ash, lava pieces, etc. noting their location;
measurement of slope angles, measurement of crater sizes by
pacing; study of drainage patterns and erosion on slopes; effects
on settlement.

Questions

1. What has been the contribution of vulcanism to the landscape of Africa?

2. With reference to examples in Africa, describe the major
forms caused by the intrusion of igneous rock.

3. With reference to examples in Africa, describe the major
landforms caused by the extrusion of igneous materials.

4. Describe and account for the mode of formation of the
following extrusive volcanic landforms:

   a strato-volcano; a caldera; a lava flow; an ash cone;
   a crater lake; a shield cone; an explosion pit.

Refer to examples in Africa.

5. Describe and account for the following intrusive features,
referring to examples in Africa:

   a batholith; a lopolith; a dyke; a sill; a laccolith.

14   Desert Landforms and Landscapes

Approximately one half of Africa's area can be classified as extremely arid, arid or semi-arid (Fig 203). One quarter of the continent's surface is occupied by the world's largest desert—the Sahara—which extends 4 800 km from the Atlantic to the Red Sea with a maximum width of almost 2000 km. The Sahara's 9·6 million km^2 would cover western Europe from the Mediterranean to the Baltic and from the Urals to the western coasts. Within this vast region are landscapes of widely differing character. Eroded volcanic rocks form the pinnacled landscapes of the Hoggar, Tibesti, Marra and other massifs, in contrast to the escarpments, basins and level plains of Cretaceous limestone or the great sand seas or ergs of Algeria, Mauritania and Libya (Fig 211). The stony deserts or reg form level monotonous surfaces of gravel and pebble mantles which merge with the bare wind-swept rocky plateaux of the hammada.

The Namib Desert forms the second largest zone of extreme aridity and consists of a 150 km-wide strip of massive sand dunes between Lüderitz and Swakopmund. To the east lies the world's largest continuous expanse of sand—the Kalahari region. These deep reddish sands are now fixed by vegetation and the dune form is now a rarity.

Other arid and semi-desert regions are on a smaller scale. The Turkana Desert, for example, west of L. Turkana in Kenya and the neighbouring Somali Desert display all the features of the larger deserts—stationary and migratory dunes, pediments and rock-strewn plains, and lake or lacustrine deposits along former lake shores.

Wind Transport and Deflation

Wind plays a much greater role in landform sculpture in deserts than in moister regions and wind or aeolian erosion, transport and deposition have free play over vast areas. The sparse vegetation cover hardly reduces the wind speed and the limited root system and low moisture content of soils prevents the particles from binding together thus exposing surface grains to removal.

When wind flows over a sand bed the speed at the base of the airflow is reduced by friction with the sand surface and the rougher this surface, the greater the friction and the lower the wind speed. A positive pressure is exerted on the windward face of sand grains and a vacuum pull or negative pressure on the leeward side. This creates a drag on the top of the grain which, if sufficiently strong, causes the grain to move. The finest dust particles are moved in suspension in the air current, medium-sized particles by saltation (irregular bouncing), and the coarsest particles by surface creep (rolling). These processes should be compared with those caused by stream flow (p. 48). Saltation is responsible for most sand movement while creep moves about a quarter of the load. The rate of creep depends on the rate of bombardment of the flying grains (Fig 204). In general most sand moves on or near to the ground surface in a zone up to 9 cm high over sand and up to 200 cm over stony surfaces. If the surface is smooth and of very fine material, such as silt, high wind speeds are necessary to start the grains moving.

The wind thus acts as a grain sorter, removing the lightest particles by deflation and leaving the heavier grains as deflation-residues. Selective deflation is responsible for the dominance of heavier minerals and surface colours in certain regions with some areas consisting almost of whitish quartz grains and others of darkish brown mica or biotite. The reddish Kalahari sands are often interrupted by zones of white sands as a result of this selective process. Very fine dusts are transported great distances in the upper atmosphere before being deposited and fine reddish dust particles from the Sahara have fallen in many areas of western Europe.

The rate of deflation depends on the degree of surface roughness and local moisture conditions. Very fine clays, silts or salts in the sand will also aid in binding the sand particles. In slightly moister semi-arid regions deflation will remove dry particles down to the sub-surface moist layer and thus form deflation hollows about a metre deep. The presence of rocks, scattered vegetation and humus in moister zones also checks the rate of deflation. Deflation lowering of the land surface is thus a slow process; only 2·4 m of surface sand has been removed over the last 2600 years in the Nile Delta region.

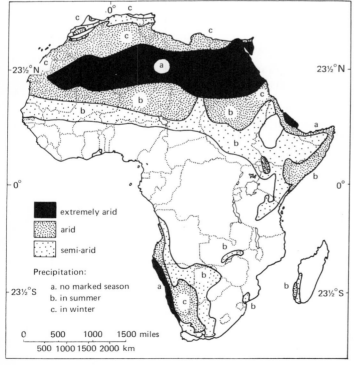

Fig 203.   The distribution of arid lands in Africa

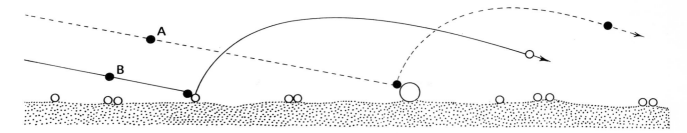

Fig 204.   The movement of sand grains by impact in arid regions. In bouncing impact, sand grain A approaches at low angle, hits larger, heavier grain or pebble, and bounces into air again. Height of bounce depends on angle of strike. In splashing impact, sand grain B strikes surface grain and splashes it into the air at a slower speed and at a lower height than bouncing grain.

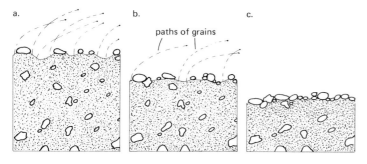

paths of grains

Fig 205.   Stages in the formation of reg or stony mantle. (a) Deflation active; (b) Deflation lessens as stones increase at surface; (c) Almost total stony surface prevents deflation

Fig 206.   Mauritania: stony mantle or reg desert caused by wind deflation. Note the angular nature of the wind-eroded stones or ventifacts

Deflation is also largely responsible for the formation of surface layers of stones some 6 cm–7 cm thick. These layers consist of coarse, angular, wind-shaped stones which overlie or are sometimes embedded in finer sands and clays. The deflation process winnows out the finer particles leaving the stones behind. Eventually the stone carpet acts as a protective layer to the loose sand beneath and deflation then ceases (Fig 205). Such deflated stone pavements or stony mantles (the gibber plains of Australia) are common in the rocky parts of the Sahara and Libyan deserts where they are called *reg* and are extensive in the northern Namib and Namaqualand. Small reg zones often lie at the foot of rocky escarpments or on the surface of alluvial fans.

Water action and moisture content also play a role in stony mantle formation. During the occasional thunderstorm the high run-off from exposed rock surfaces will sort the stone particles. Sub-surface stones will also rise through finer material by drying and wetting. Sand grains expand when wet and force stones to rise; with increasing dryness dry sand particles acting under gravity fill in the spaces beneath the risen stones and prevent them from sinking back to their former position. Eventually the stones will be concentrated at the surface (see experiment p. 113).

Fig 207. Chad, northwestern edge of the Tibesti, near Bardai: a sandstone pediplain above which rises an isolated inselberg whose summit is protected by a cap of volcanic lava

Wind Erosion and Landforms

The sands of Africa's deserts are derived from several sources. Some sands are the product of the erosion of sandstone (itself originating in the cementation of sand grains in an ancient desert) such as the Nubian sandstone of the Nile region or the Stormberg sandstones of Namibia. The breakdown of granites and schists are another source. Large areas of sand in the coastal regions of the Namib, Mauritania, Senegal and Morocco originate from former extensive beach deposits which were once exposed when sea levels were lower. Another important source lies in the vast tracts of alluvial sands laid down in past pluvial periods on river and lake beds. Such beds are extensive in the Great Western Erg of Algeria, between the Niger and Timbuktu in Mali, and in many parts of the Kalahari.

Abrasion and deflation in deserts destroy or remove the low density minerals such as mica while the high density resistant minerals such as quartz remain. Under moister conditions, such quartz grains would be protected by a thin enveloping layer of capillary moisture which migrates from lower depths but in deserts such a layer is normally absent. The constant jostling of the jagged quartz particles slowly abrades them until they become smooth and rounded and look like grains of millet seed. In the Kalahari, where there is often no capillary moisture, the grain surfaces become highly polished and produce high pitched sounds when disturbed; they are appropriately called the roaring sands.

The millet-seed grains are effective tools of erosion. Rocks lying in the path of such saltating sand grains are regularly bombarded and sand blasted, maximum erosion occurring within the 2 m-high saltation zone. The rate of erosion depends on wind velocity, grain hardness, and the softness and structural weakness of the rock. Pebbles and boulders lying in the path of the wind-borne sand will have their windward faces worn flat and such ventifacts and wind-faceted stones are common in the reg. Lighter stones may have their positions altered by a change in wind direction (Fig 208) or by sheet wash which results in two or more faceted sides, the three-sided stone being termed a dreikanter.

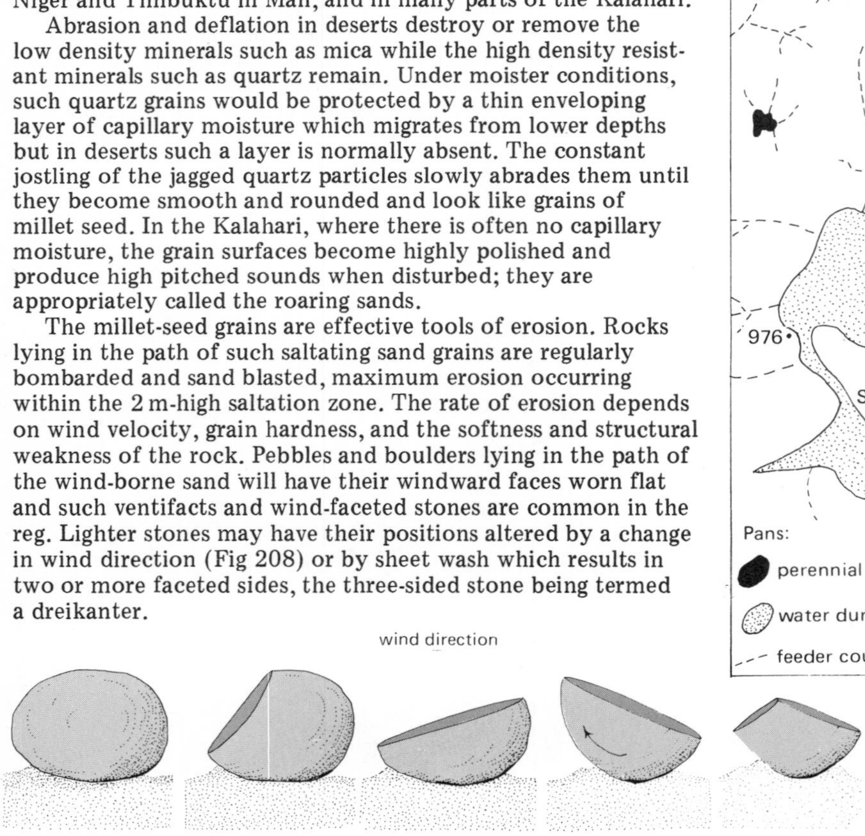

wind direction

Fig 208. The formation of ventifacts by sand blasting

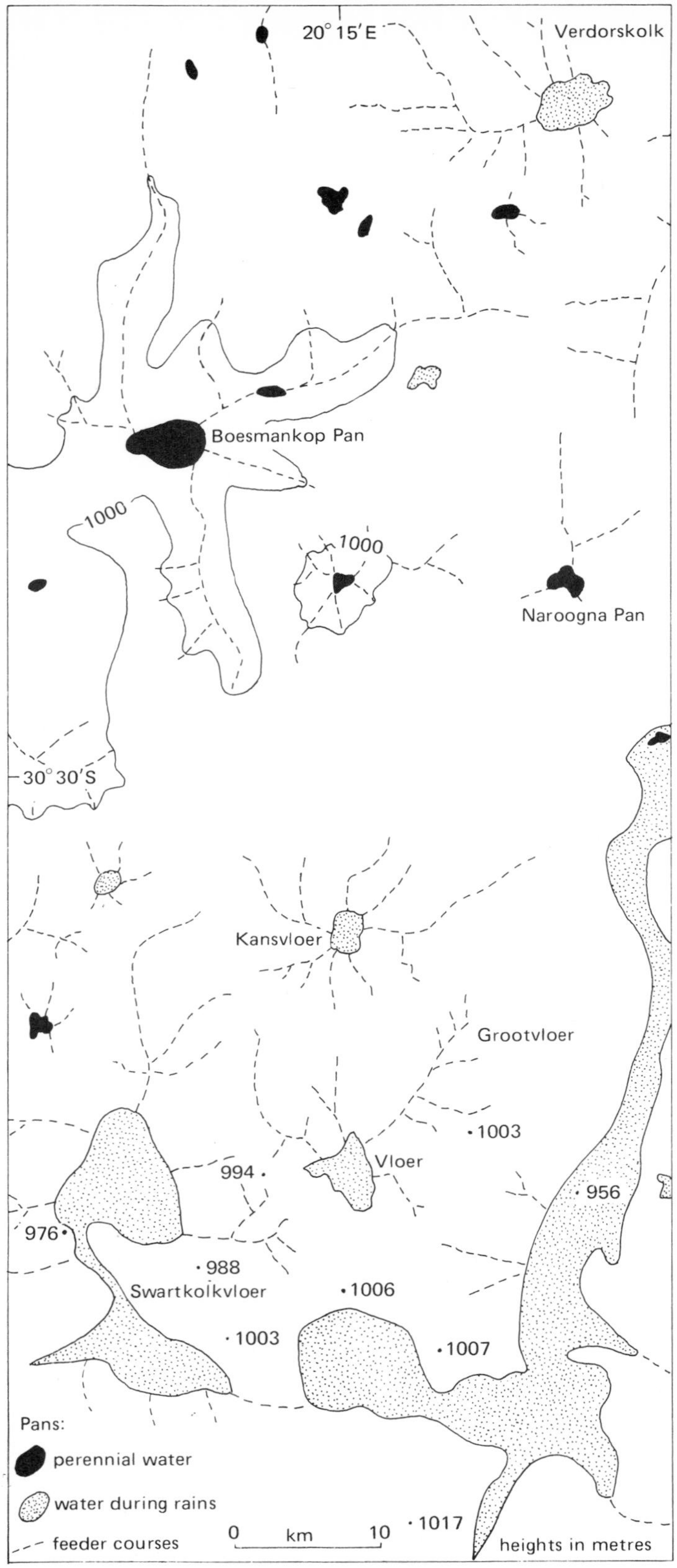

Pans:

⬛ perennial water

⬭ water during rains

- - - feeder courses

0 km 10

Fig 209. Examples of pans in the western Cape Province of South Africa. The region lies about 150 km south of the Orange River in Bushmanland

heights in metres

20° 15′ E

Verdorskolk

Boesmankop Pan

1000

1000

Naroogna Pan

30° 30′ S

Kansvloer

Grootvloer

·1003

Vloer

·956

994·

976·

·988

·1006

Swartkoîkvloer

·1003

·1007

·1017

Where the rock contains exposed bands of hard and soft rock, differential sand-blasting leaves the hard bands jutting out. Zones of structural weakness are chiselled out to form deep furrows and ridges elongated with the wind direction termed yardangs. Most yardangs are less than 3 m high but some exceeding 20 m in height and up to 1 km long exist near Borku in the northern Chad region.

If a hard cap of rock overlies softer beds the erosive processes may break through the cap to attack the softer rocks and create zeugen—table-like masses of hard rock resting on pedestals of softer rock. Similar structures but of different origin are the gour of the Sahara and Sinai deserts. Gour are mushroom-like forms of uniform rock in which the lower section has been eroded into a pedestal while the tabular section remains above the general level of erosion. Rock columns such as the Finger of God pedestal in Namibia are formed in a similar manner. When the pedestal is finally worn through the upper part of the column will collapse.

Larger scale erosive features of deserts include grooves, pans and basins. Grooves are long, narrow, wind-eroded scars on rock surfaces aligned with the prevailing wind direction. They are 3 m–15 m deep, 500 m–1 km long, and from 500 m–2 km apart. They are common along the northern edge of the Great Eastern Erg, along the southern shores of the Algerian chotts, in the northern Sudan, and to the southeast of the Tibesti Mountains.

Pans (playas, U.S.A.; *sebka* in Arabic; *chotts* in the Atlas) are very common in the semi-arid regions fringing the Sahara and in the Kalahari. They are circular depressions in generally level areas and vary in size from a few square metres to over a thousand hectares. Some are slight depressions, others deep hollows 5 m–10 m below the general surface. The floors of pans are usually flat and composed of fine-textured clays. Water collects in the pans from time to time but, failing to drain through the impermeable clay floor, soon evaporates. There are many theories regarding pan formation. The pans of the semi-arid central Orange basin in South Africa were once believed to be remnants of old abandoned courses of the Orange and Vaal rivers but there is little proof of this. The most likely explanation is that water collects in small hollows or in a stream course blocked by rocks; long periods of flooding and drying out increase the chemical weathering of the floor of the hollow especially in soft mudstones and shales and this forms an impervious fine-grained floor. Animals drinking at the pan compact its edges, break down the soil structure, and carry away small quantities of soil. As the surface dries out the wind picks up the fine grains and deepens and enlarges the depression.

Some pans contain fresh water, others salts such as sodium chloride or carbonate derived from surrounding rocks and saline soils. The pans are centres of centripetal drainage systems and have no outlet so that continued evaporation leaves successive layers of salt deposits. Salt pans or salinas are common in eastern, central and southern Africa, the best known being L. Magadi in Kenya which yields commercial deposits of sodium and salt, and L. Natron in Tanzania. Very large pans are located in Egypt's western deserts although some may be due to local downwarping. Some of the oases in Libya and the Qatara Depression are deflation hollows which have reached the water-table level.

Fig 210. Algeria: in this region of the Hoggar Mountains, the crystalline base of Africa rises to the surface to form a massif of granite and gneiss with a general height of 2 000m. Much of the surface has been buried beneath lava flows 180 to 210m thick. The lava has been weathered into towers and turrets and here, in the Assekrem, there are about 300 of these volcanic pinnacles

Windblown Depositional Features

About one quarter to a third of the world's desert surfaces are composed of aeolian (wind-blown) sands forming vast sand seas or ergs. Ergs vary in size, the smallest being 125 km² and the largest, the Great Eastern Erg of Algeria, is almost the size of France. Ground photographs of such sand seas often give the impression of a chaotic landscape but in reality erg sand formations are arranged in a complex but orderly pattern of sand dunes of various sizes. Not all sandy regions develop dunes, however. The Selima sand sheet near the Sudan-Egyptian border, for example, is a very smooth layer of sand about 30 cm thick covering several thousand square kilometres and has an average gradient of only 1:1500. In some areas of the Kalahari loose sand surfaces have no dunes, but merely irregular low undulations no more than 5 m high.

Usually, however, the dominant landforms of the ergs are bedforms—shaped mounds of sand of varying sizes which form complex but regularly repeated patterns. Close similarities exist between desert bedforms controlled by windflow—ripple, sand-dune and wave patterns—and sand-forms on the beds of rivers and seas controlled by water flow. Wind speed controls the volume of sand movement and thus the size of bedforms, wind flow patterns control their shape, and wind direction their alignment. Spacing between sand features is controlled by wind turbulence (updraughts and downdraughts) and by the twisting or corkscrew motion of ground-level air currents (Fig 212).

The most common bedform is the wave form. The smallest wave forms are the ripples which may be caused by either small-scale turbulence in the airflow base (aerodynamic ripples) or by the impact of bombarding sand grains (impact ripples).

These smaller forms have wavelengths (the distance between their crests) of 0·5 cm–250 cm and heights up to 100 cm. Dunes have wavelengths between 3 m and 600 m and heights up to 100 m, while the large wave forms, draa, vary in height from 20 m to 450 m and have wavelengths of between 300 m and 5500 m.

The size of these features is related to the size of the sand grains, the coarser the grain, the greater the wind speed needed to move them and the larger the resulting feature. Sand dunes are formed where a lowering of wind speed causes deposition of sand, for example, in the lee of rocks or a bush, or where wind currents converge, or where the surface is very coarsely grained. A dune will grow in the depositional area, termed a sand patch, by the piling of sand grains, its height governed by the wind speed which is greatest in the upper section of the air current. When the dune height reaches a certain wind-speed level the upwind side will begin to erode. Sand grains will migrate up the windward side, the lee side becomes increasingly unstable until downward slumping causes a slip face with an angle of between 33° and 34°. The constant movement of grains from the windward to the slip face causes the slow migration of the dune. Height, width and weight of the dune control its speed of movement, 9 m a year being observed in the Sahara and up to 450 m a year in the Namib.

Draa are giant dunes with similar cross-sections and shape to dunes. Their movement is controlled exactly like dunes but is much slower, about 50 cm a year or less. Draa and other dunes observed in Egypt, Algeria and the Sudan usually occur in regularly spaced ridges caused by wind eddy patterns. It is quite common to see all three wave bedforms together with smaller dunes migrating along the backs of draa and ripples moving along the backs of the dunes.

Fig 211. Algeria, the Great Western Erg between Bechar, and Reggane: ripples and migratory dunes are seen in the foreground with draa on the skyline

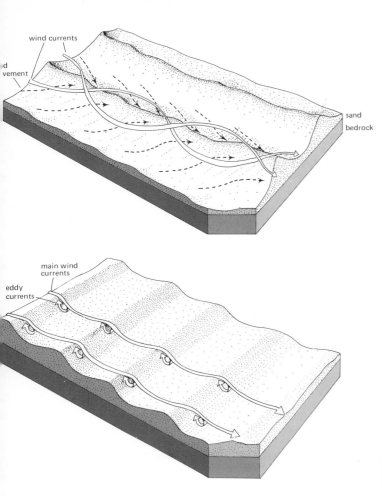

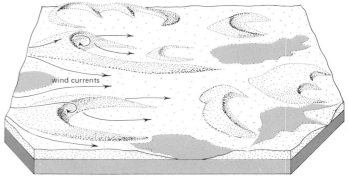

Where winds blow steadily from one direction both arms of the barchan may be elongated. Some barchans have only one elongated arm possibly due to a greater sand supply on that side or to a seasonal change in the wind pattern. Neighbouring barchans sometimes have adjacent arms elongated, a phenomenon not yet fully understood. Barchans are common in Algeria and Chad and in parts of the Namib.

Fig 213. Some shapes adopted by barchans

Fig 214. Namibia, the Namib Desert: the migration of linear dunes is halted by the Kuiseb River; to the north of the river the sand is in sheet form and is crossed by several wadis. Note the sand spits and sand shoals along the coast

Fig 212. Longitudinal and transverse dune formation
Longitudinal dunes: The sand is drawn into a long ridge by the corkscrew currents of wind.
Transverse dunes: The sand is formed into low waves transverse to the wind direction by rotating eddy currents.

Dune Forms and Patterns

The simplest form of dune is the tamarisk mound (*nebka* or *rebdou*), formed where sand piles against a small tree or boulder. The wind rises over this obstacle and deposits sand in the wind shadow. On rough, flat sand surfaces wind deposits sand in very low ridges with long wavelengths and with no slip faces, called zibar. Lee dunes form where the air stream breaks to flow round an object such as a small inselberg; lee or echo dunes form in the comparatively calm air immediately behind the object and may stretch for up to 3 km. Eddy currents also cause the deposition of similar linear dunes parallel to the edge of long ridges. In the clefts and hollows of rocky ridges small dunes may be seen migrating up one side of the ridge and down the other.

Barchans form on hard rock surfaces and where sand supply is low. A sand patch on the rock surface lowers the wind speed at that point by friction and sand is deposited on the patch. Wind speeds will be higher at the sides of the patch on the bare rock surface. Thus the middle of the sand pile will grow while long tapering arms pointing downwind will develop at the sides. Barchans have been observed to migrate at speeds of 24 km a year in Egypt and become stabilised or fixed dunes when they reach a height of about 35 m and a width of 380 m.

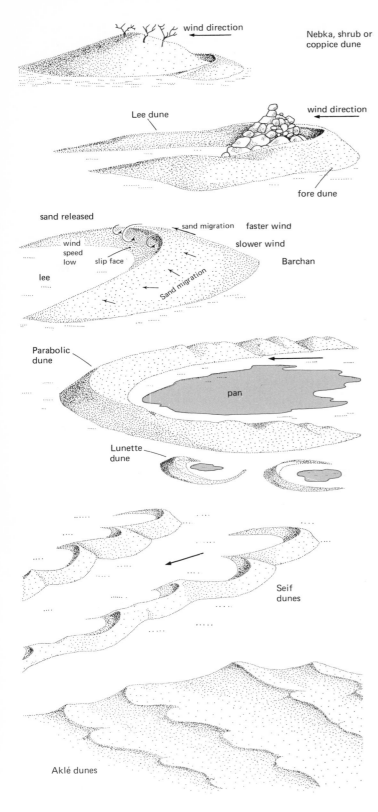

Fig 215. Some of the major types of dune formations

Where a ridge of sand lies across the wind direction, the wind force may blow out sand to form a huge hollow within pointed arms. The result is a dune like a barchan but the horns point upwind instead of down. These parabolic or U-dunes have a long gentle windward slope and a steep lee slip-face like the barchans. Occasionally the centre part is completely blown away to leave the arms only.

The seif dune (*seif* is Arabic for a curved sword), is common in regions where winds blow steadily from two or three directions during the year. Seifs are probably derived from barchans which form with the prevailing wind while the elongation of one arm is caused by a weaker but persistent wind, possibly seasonal, from another direction. Seif dunes are usually straight and long with occasional curved arms branching from the main line; the crests are sharp and the slopes on either side are approximately 20°.

Where lines of dunes meet or cross, huge pyramid-like dunes called rhourds or sand mountains are formed. Rhourds may also be caused by concentrated updraughts of air due to local convection currents which draw in loose sand and pile it up. Rhourds are common in the Great Eastern Erg.

A frequent depositional feature of the Kalahari and Senegal sand surfaces is the lunette dune composed of fine sand or silt blown from a pan and deposited downwind to form a crescent-shaped ridge along the shore line. Lunettes have rounded crests and long gentle lee slopes.

In addition to the sand seas of migratory dunes, there are wide zones of fossilised dunes fringing the present deserts in the Sahel and Kalahari. These zones of dead dunes were created during phases of greater aridity–the interpluvial periods of the Pleistocene—when dry conditions were more extensive. One of the largest zones of fixed dunes lies in the Hausaland region of northern Nigeria. Here massed dune ridges are elongated in an ENE–WSW direction, their outlines modified by erosion and their surfaces now fixed by vegetation. The dunes are similar to the existing active linear dunes of the Libyan Desert. These fossilised dune formations lie south of the 150 mm isohyet although there are some mobile dunes still existing in the same region (Fig 216). Ancient dune fields are also located on the flanks of the lower Senegal and to the northeast of L. Chad. The problems of dating such ancient ergs are still a matter for research.

Water Action in Arid Regions

The low rainfalls and high evaporation rates of arid regions prevent the formation of permanent rivers. Rivers such as the Nile and the Niger derive their waters from regions outside the deserts. Some rivers such as the Ewas Ng'iro (or Uaso Nyiro) in Kenya which ends in the Lorian Swamp or the Awash of Ethiopia which ends in the Danikil Desert, rise in rainy highlands but never reach the sea, their flow being reduced by evaporation and percolation. Deserts and semi-deserts are generally regions of internal drainage, the water draining into enclosed basins which may contain shallow lakes or large swamps, e.g. L. Chad, the Lorian Swamp in Kenya, the Chobe Swamp of northern Botswana, the Etosha Pan and the Okavango Basin of Namibia and northwestern Botswana.

Despite low rainfalls many of the landforms of arid regions have been formed by flowing water and some geomorphologists believe water to be more important than wind in the formation of desert landforms. Even the driest regions receive occasional rainstorms which cause rapid run-off and play a major role in shaping the landscape. For example, 200 mm fell in a single day at Port Etienne, Mauritania in 1913 and 88 mm in one day at Villa Cisneros, Spanish Sahara. On 7 December 1960, some 16 mm of rain fell on the Tadmaït Plateau in central Algeria,

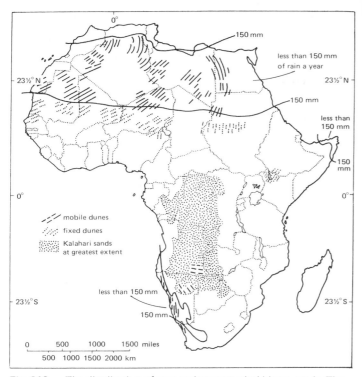

Fig. 216. The distribution of past and present wind-blown sands. The Kalahari Sands extended into the Congo basin during a dry interpluvial period. Note the general alignment of the dunes in the Sahara region with the prevailing wind directions

Fig 217. Algeria: aerial photograph of star dunes or rhourds in the Great Eastern Erg. The height of these dunes varies between 180 and 240m and the diameter varies between 2.5 and 3.5 km.

the driest part of the Sahara, causing run-off streams to flow at the rate of $120 \, \text{m}^3 \text{s}^{-1}$, each litre of water carrying $1 \cdot 7 \, \text{g}$ of weathered rock and sand. Such streams and flash floods are even more common in the semi-desert regions and maintain their flow and erosive effects for longer periods and over longer distances.

Many of the landforms in today's deserts have undoubtedly been formed during past periods of heavy rainfall. These pluvial periods of widespread and intense rainfall continued long enough to have a lasting effect on landforms and landscapes which were then preserved during later dry periods. About one million years ago during the Pleistocene period, a pattern of pluvial and dry interpluvial periods began which had a lasting effect on the surface features of the Sahara, the Kalahari and the Namib deserts. In eastern and southern Africa the last major pluvial period lasted some 22 000 years (from about 38 000 to 60 000 years ago) with another slightly moister period from 12 000 to 20 000 years ago. Over the last ten thousand years the Sahara has experienced a dry period lasting 2500 years, then a moist period 3000 years long, followed by the present intensely arid period. During the moist periods open tropical woodland flourished in the western Sahara, savanna flanked the Nile, and vegetation zones now on the edges of the deserts extended some 250 km nearer to the present desert cores. In the Tassili, Adrar, Tibesti, Aïr and Uweinat highlands man has left pictorial records of hunting scenes depicting animals now found only in savanna regions—elephants, water buffaloes, antelopes, ostriches, giraffes.

There is also much evidence to show that pluvial lakes occupied depressions now dry and that existing lakes, playas and salinas were much more extensive with networks of contributary rivers. Strand-lines at higher elevations than present beach surfaces indicate that many lakes were formerly much

larger (p. 70). The strand-lines above the Kenya lakes in the Rift Valley also reveal evidence of the homes of fishing people. Mud deposits have yielded pollens proving the existence of oak, pine, wild olive and cypress in the Sahara. Fossil soils formed under moister conditions have been revealed along the coasts of Senegal and Mauritania, in the Saharan plateaux and along the Nile terraces. The deep chemical weathering of the exposed crystalline rocks of deserts in Africa could only have occurred during such wetter times. The domed inselberg (p. 29) may have formed during a former summer-rainfall tropical climatic regime.

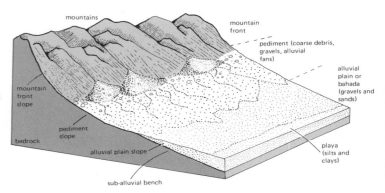

Fig 218. The major landforms associated with arid and semi-arid regions

111

Desert Landforms Caused by Running Water

Deserts therefore contain many landforms created by running water. In mountainous regions vertical fluvial erosion produces deep V-shaped valleys and ravines separated by rocky, sharply defined interfluves—typical badlands topography. Such a region is the Acacus area of the Fezzan in Libya. Here, limestone cliffs fall steeply westwards and are gashed and crisscrossed by narrow gorges and tortuous defiles. The granite of the region is eventually broken down into rocky heaps called gargaf.

In the deserts the processes which cause the parallel retreat of slopes and the emergence of the pediplain are often clearly seen. The debris or detritus from the uplands (the waxing slope) is deposited at the foot of the mountain front (the pediment) where the occasional flow of streams is checked by the abrupt change in gradient and where the water seeps rapidly into the desert surface. Such moisture seepage may cause slumping as, for example, along the Tadmaït escarpments in Algeria. At clefts in the escarpment face stream courses may join and discharge their gravel, sand and mud to form an alluvial fan—a low, half cone-shaped debris accumulation whose size and form is controlled by the rate of debris supply and the channel network flowing over its surface. Alluvial fans are common desert forms especially where plenty of debris has accumulated on upper surfaces and where channels focus on to flat lowlands. The debris is graded in size, the finest particles being washed to the base of the fan. The radial diverging stream channels cut the surface of the fan into dendritic and braided patterns, these patterns constantly changing with each water discharge. The largest fans have radii up to 10 km but slopes rarely exceed 10°. Subsidiary fans may also form (Fig 219) and several fans may unite in a continuous plain termed a piedmont alluvial plain.

Fig 219.   The development of alluvial fans and a piedmont alluvial plain.

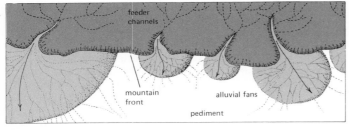

(a) Alluvial fans forming at discharge points of rivers from the

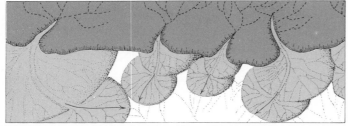

(b) Piedmont alluvial plain forming at base of mountain front by the merging of alluvial fans

Occasionally the water velocity is sufficient to carry very fine-grained clastics (clays and silts) and non-clastic minerals (salts) on to the flat desert surface and into depressions to form temporary lakes or pans. The Chott el Djerid is a large pan which becomes a shallow lake in winter but in summer water is confined to small holes on the lake bed surface called aïoun. Pans also display low cliffs and terraces caused by water abrasion, overflow channels, beaches above the present water line, and spits and sandbars along their shores.

Not all stream channels on the desert surface are shallow. Intermittent water flows may be confined to a single channel on several occasions. Since vertical downcutting exceeds lateral erosion in arid regions because of the general lack of surface water, a gorge-like channel with steep rocky sides called a wadi is formed (Fig 220). Each flow of water deposits debris but wind deflation will keep the depth of the wadi floor below the general desert surface. The largest wadis are partly due to river erosion during past pluvial periods. The Wadi Zemzem in Libya,

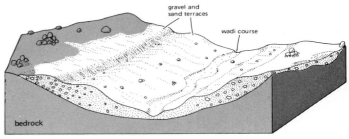

a. Wadi developed on sand and gravel surface

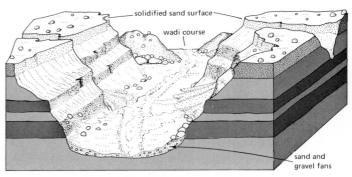

b. Wadi eroded into rocks of varying hardness

Fig 220.   Block diagrams illustrating two types of wadi

for example, is some 160 km long and leads into the Tauorga Marshes. It has several tributary wadis, the largest of which are the Wadi Tamet el Mgenes and the Wadi Bey el Kebir. Even today in times of flood the combined waters of these tributaries may be large enough to reach the Gulf of Sirte on the Mediterranean coast.

Over many thousands of years of deposition, the wadis and other channels become choked with debris. The whole land surface adjacent to the mountains becomes covered with a vast mantle of detritus and is termed a bahada which gradually rises higher until it encroaches on the alluvial fans and the footslopes of the mountain scarps (Fig 222). Outliers are gradually buried and appear like islands in a sea of sand; spurs along the mountain front jut into the bahada and embayments in the scarp face are gradually infilled with debris. When the bahada debris has engulfed the remnant mountains a level, almost featureless, rock-strewn plain emerges. Such bahadas flank the mountain masses of Jebel es Soda, the Asuad Highlands, and the northern Tibesti in Libya.

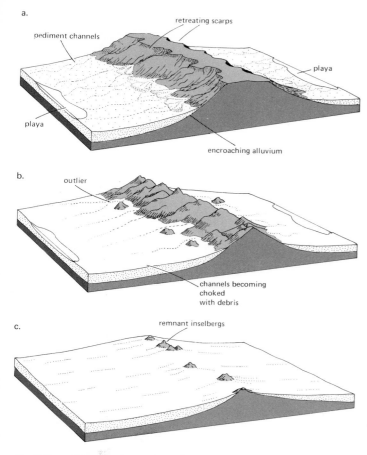

a.

pediment channels

retreating scarps

playa

playa

encroaching alluvium

b.

outlier

channels becoming
choked
with debris

c.

remnant inselbergs

Fig 221. Stages in the erosion of a mountain range and the development of a rock-strewn plain in an arid region

Experiment

Stony mantle formation: Obtain a box about 25–30 cm deep and fill it to about 3 or 4 cm from the top with dry sand and small angular stones not more than 2 cm long (gravel chips would do). Place the gravel chips near the base of the sand making sure they are well spaced. Then water the surface of the sand regularly allowing it to dry out before watering again. Once a day will be sufficient. After about a fortnight to three weeks the stone chips should appear at the surface depending on the number of times of watering and drying. Explain your experiment.

Fieldwork

During the dry season in tropical areas conditions similar to those in semi-arid and arid regions may be experienced. Students should look for evidence of exfoliation and pressure release features on rock surfaces. Vertical wooden boards can be set in the ground, their alignment at right angles to prevailing winds so that sand from fields builds up against the board. A series of such low boards will give some indication of the rate of removal of grains from the ground surface. Is the removal rate the same for areas with a slight vegetation cover? Are there any boulders or stones which show evidence of sand-blasting? Examine the surface of rocks closely. Are there any loose grains in the microfissures? Are there mineral particles in the fissures? What are the effects of sand-blasting on buildings? Make notes on your observations.

Questions

1. Describe the processes involved in wind erosion and the major features produced by these processes.

2. Describe and explain the mode of formation of the following desert landforms:

   pans or playas; rock pedestals; yardangs; erosion grooves; deflation hollows.

3. Describe the following desert features and explain their mode of formation:

   dreikanter; gibber plain; sebka; hammada.

4. Write an essay entitled 'The Depositional Forms of the Deserts of Africa'.

5. Describe and account for the following depositional features in desert regions:

   ripples; tamarisk mounds; barchans; seifs; rhourds; draa; U-dunes; lee dunes.

6. 'The landforms of today's deserts are due more to the action of flowing water than to wind or isolation'. Discuss this statement.

7. What evidence is there that today's deserts in Africa were once subject to prolonged rainy periods?

8. Describe and account for the following desert landforms:

   alluvial fans; wadis; bahadas.

Fig 222. Algeria: the Wadi Tamanrasset in the Hoggar Mountains

15   Coastal Landscapes

Africa is remarkably compact in shape despite its vast size and its coastline is a generally smooth one broken only by a few small estuaries and limited stretches of deep inlets. River mouths partially closed by sand-spits, shallow lagoons enclosed by coral reef and sand-bars, or low dune coasts backed by hills or mountain ranges are the more usual features.

The character of the coastline at the junction line of water, land and air is controlled by several factors: the agents of marine erosion—ocean currents and waves; the resistance and type of mainland rock; the dip of the rock strata; the depression or elevation of the land relative to sea-level; the deposition of wind-blown or water-transported sands; or the growth of coral reef under special conditions.

Wave Action

Waves are the chief agent of marine erosion and deposition. They are caused by the brushing or drag action of wind on the water surface transferring atmospheric energy to the sea surface; the action produces an energy movement which passes through the water. The movement is similar to wind currents passing through grassland—the water hardly moves just as the grass remains rooted as the energy waves pass through it. Thus a floating object will hardly move forward with a passing wave but will rather rise from wave trough to crest then fall with the next trough (Fig 223).

The size of the wave and the work it can do depends on the amount of energy transferred to the water by the wind and on the wind's duration. Small waves observed on pans or water holes appear as mere surface ripples on the windward side where the wind is only just beginning to skim the surface, but they become much larger on the opposite side where the energy accumulation is greatest. The wider the lake, the longer the distance over which the wind travels (the fetch) and the larger the waves at a given wind speed. Out at sea under a strong wind rollers may form rising to 15 m between crest and trough (the amplitude) with wave lengths between successive crests exceeding 1000 m. If the wind stops, such rollers will continue their momentum as swell. Along coasts with strong winds and with long sea approaches, giant sea swells produce huge destructive waves, for example, during winter along the coasts of the Cape Peninsula, Algeria and Morocco. On coasts in West and East Africa huge breakers may be seen crashing against reefs and sand-bars while within the lagoons the beaches are lapped by gentle waves caused by the short fetch.

When waves run into shallow water their circular motion is broken by the inclined sea bed (Fig 224). The largest waves will break further out to sea but smaller waves will break nearer the shore. Eventually the forward movement is checked, the wave steepens and becomes top-heavy and the wave becomes a wave of translation, collapsing as surf or breakers. Practically all the wave's energy is expended in this surf zone, the base of

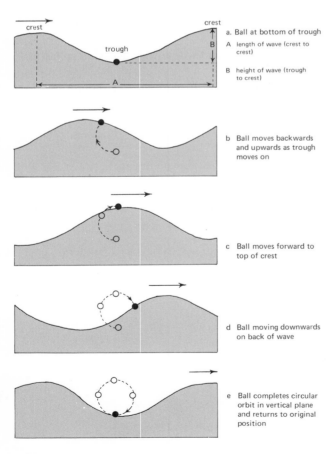

Fig 223. The motion of a table-tennis ball as a wave passes over the sea surface

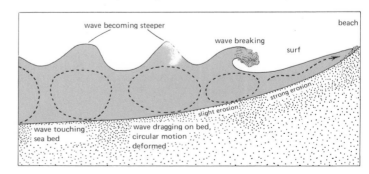

Fig 224. The circular motion of a wave being broken as it enters shallow water

which is usually less than 2 m deep. Air trapped within the water compresses and expands to create foam.

Two types of waves occur (Fig 225). More gentle waves, breaking at the rate of about one every 10 to 12 seconds, have a more powerful forward thrust, their swash being greater than their backwash—a constructive beach-building action. Larger waves occurring one every 5 seconds are usually destructive since their backwash is more powerful than the swash due to the vertical fall and combing action of the wave. The nature of the beach also affects these constructive and destructive actions—a shingle beach composed of shelly fragments and rocks of more than 2 mm size allows rapid downward percolation minimising backwash; a sandy beach of quartz, feldspar and mica fragments of less than 0·1 mm may lose much material to backwash. Steep beaches are also more likely to suffer loss through backwash than more gently inclined beaches.

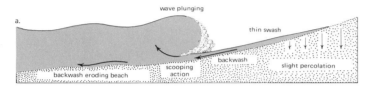

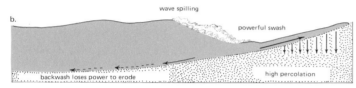

Fig 225. (a) Destructive wave; (b) constructive wave

Where waves approach the shore at an angle the part of the wave nearest the shore will come into contact with the bottom first and will move forward a shorter distance than the same wave further off-shore. The crest of the wave thus alters direction—an action termed wave refraction. On an identical

coastline with bays and headlands, wave energy is concentrated on the headlands due to refraction (Fig 227) while in the bays the energy is more widely distributed. The headlands will have loose debris removed and beaches are unlikely to occur, while the embayments nearly always possess sand or shingle beaches.

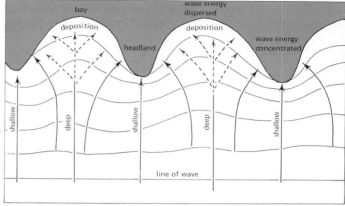

Fig 227. Wave refraction caused by jutting headlands helps to concentrate erosive power of waves on headlands. In the deeper bays the energy of the waves is dissipated and deposition rather than erosion is more likely

Currents may be caused by tidal variations especially where the water passes through a narrow channel in a sand-bar, but also they may be due to refraction of waves at headlands. This causes water to build up at the headland and flow to the lower levels in the bays. Rip currents are concentrated backwash flowing seawards beneath and through the advancing waves to produce a strong undertow. They are particularly strong off the coasts of Natal, the Transkei, Ghana and the Ivory Coast and are dangerous for bathers. Currents are responsible for small-scale beach features such as sand ripples and sand waves and their action is comparable with wind effects in deserts (see p. 108).

Fig 226. Kenya: Tiwi Beach about 20 km south of Mombasa. The low cliffs are old coral raised during the Pleistocene period to form a platform. Marine erosion has produced a wave-cut coral platform (foreground). The sea is attacking the base of the low cliffs to excavate caves — note the notch indicating low and high tide erosion levels. The beach (left), is largely composed of coral particles

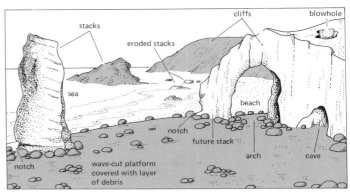

Fig 228. (Left) Namibia, the Bogenfels Arch situated on the rocky coast of the southern Namib desert: the arch is about 55m above sea-level and is composed of massive dolomite. This coast is noted for its steep cliffs, stacks and arches

Chemical processes also influence cliff formation and retreat, particularly in coral or limestone regions. Both mechanical and chemical weathering are highly selective, the softer rocks being removed and the harder materials standing out as resistant fore-lands and promontaries. The sea pounds into joints and widens them by chemical and mechanical erosion to form caves, arches and finally stacks (Fig 230).

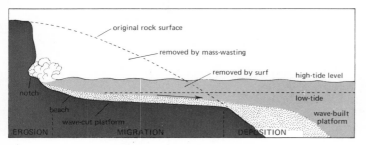

Fig 230. Erosional features of a rocky coast. The cave will develop into an arch, the arch will collapse to produce a stack. The angle of the cliffs depends upon the dip of the rock strata

Wave Erosion and Coastal Features

Waves are very effective agents of erosion. When a wave's forward motion is suddenly checked by a rocky promontory of cliff face its energy is converted into hydraulic pressure. The water is flung to great heights on impact and the rock is subject to tremendous pounding. Huge storm waves like those which batter the cape coasts of South Africa can produce pressures of up to 3000 kg m^{-2}. Air in joints and fissures is suddenly compressed and then quickly released as the wave falls away. Rock surfaces continually subjected to this pressure-release action are under powerful stress; pieces are prised loose and fall under gravity and are then subject to abrasion by being jostled against one another until they are finally reduced to individual mineral grains. Larger fragments are hurled by the waves agains the cliff face and hasten erosion. Such erosive activities are greatest between high and low water marks with honeycomb weathering occurring above the high water mark by higher flung particles. Cliffs will be undercut and faces will collapse. The collapsed debris will protect the cliff base for some time but, once removed, the undercutting will begin again and a wave-cut platform or shore platform will emerge (Fig 229). The materials removed from the cliff face are dragged over this platform, abrading it and themselves and then are deposited immediately offshore to form a wave-built platform. The less compact the rock, the shorter the time taken for such processes.

Cliffs along Africa's coastline often form prominent breaks and easily recognised landmarks in an otherwise low sandy coast. The coast of West Africa, for example, is generally low-lying but is occasionally broken by steep cliffed headlands as at capes Three Points, Palmas, and Verde. Cliffs occur in the Freetown Peninsula, at Cape Mount near Monrovia in Liberia, in the sandstones of Gambia and near Mt Cameroon. Here the basaltic lava flows of Cameroon volcano (Fig 197) form cliffed head-lands separated by embayments whose beaches are of black basaltic sands. The basaltic cliffs are being rapidly eroded by chemical action which is three to four times swifter in sea water than in fresh. Impressive cliffs are a feature of the Cape Peninsula coasts of South Africa where folded truncated anti-clines end in sheer cliffs, e.g. Cape Hangklip, Cape Point and the Marine Drive cliffs of Table Mountain Sandstone east of Cape Town. There are several examples of wave-cut platforms in this region and further along the coast at Hermanus.

Differential erosion has formed a promontory and bay coast with numerous stacks along the coast of Ghana between Bishop's Boys' School and Accra, at Sekondi and between Accra and Cape Three Points. The Entebbe Peninsula region of L. Victoria in Uganda has several small resistant headlands which jut out into the lake and which are flanked by small bays in which sand is being deposited, e.g. near Entebbe itself and also at Kibanga. In parts the lake waters have breached the granite rocks along the shore and excavated embayments in the softer rock further inland. The small islands off-shore here are rather like low stacks but are in reality the remnants of former headlands (or spurs) now submerged.

The rocky coast of the southern Namib also shows the effects of differential erosion by the sea. Where older crystalline rocks of varying hardness are exposed the coast is rocky and indented but where gneiss outcrops cliffs are the common feature. In other parts quartzite rocks form steep cliffs while softer shales have been eroded into shallow bays.

Fig 229. Erosional and depositional processes along the sea coast: notching of the cliff base and the formation of a wave-cut and a wave-built platform

Wave Action and Depositional Features

Beaches may be of two types—tidal or tideless. The tidal range of the sea around Africa's coasts varies considerably from the almost tideless Mediterranean to 1·5 m between high and low tide in the Forcados estuary, 2 m in the Bonny estuary and as high as 7 m at Boké in Guinea. Along the Mediterranean shores of Africa the waterline position hardly varies and at this line a pronounced step in the beach sand is caused. The high line of the step coincides with the upper limit of swash, the low line of the step with the limit of backwash. Constant addition of sand to the slope causes it to migrate slowly seaward, its height governed by the size of the waves. Such narrow stepped beaches are typical of the Algerian and Tunisian coasts.

Tidal beaches have much wider foreshores with a very gentle gradient; the greater the tidal range, the wider the foreshore. Sand beaches have a gradient less than 1:100 while shingle beaches are 1 in 7 or steeper. The beaches may have no particular surface pattern, e.g. on the soft coral beaches of East Africa, or they may have slight surface swells about 1·5 m high called fulls which are aligned at right angles to the wave approach direction. The back slopes of these fulls are usually rippled.

When waves approach a beach obliquely they are able to transport vast quantities of sand and pebbles along the beach by longshore drift. A breaking wave washes up the beach at an angle taking with it sand and small stones which return with

Fig 231.   South Africa, the Cape Peninsula: the horizontal strata of Table Mountain Sandstone are being attacked chemically and mechanically by the sea. The results are steep vertical cliffs, jutting promontaries and an emergent stack in the foreground

the backwash under gravity down the beach slope and at right angles to its alignment (Fig 232). Particles are transported in zig-zag fashion and can only be halted by some obstruction such as a man-made wall or groyne which traps the sand to form a pocket beach, e.g. along the Natal coast near Durban and on the Bathurst Peninsula in Gambia (Fig 239). Where the beach terminates at an estuary or headland the longshore drift process

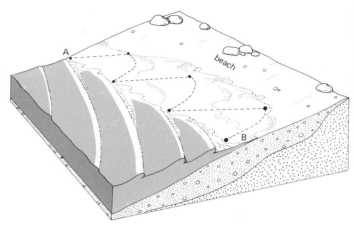

Fig 232.   Longshore drift: The sand particle at A is carried at an angle up the beach, returns in a curve by gravity, and migrates to B by repetition of the process

Fig 233.   Nigeria: a long barrier beach backed by lagoons, near Lagos

carries material onwards, depositing it as a spit attached to the headland. These may be sand-spits or shingle spits. Sometimes the end of the spit is curved landwards due to oblique waves advancing along the spit and swinging round the end of it (Figs 234 and 239).

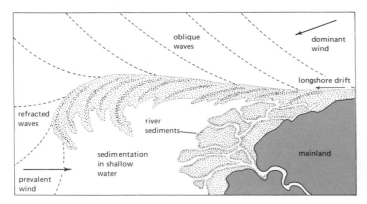

Fig 234.   The development of a hooked spit caused by the migration of sand due to longshore drift and the refraction of waves round the end of the spit. The backwaters of the spit are gradually being filled in with river sediments

Sand-bars derive their material from the seabed rather than from longshore drift and are elongated ridges of shingle or sand running roughly parallel to the coast. Sand-bars develop where surf causes sand grains to migrate seaward (Fig 235). Constant repetition concentrates the sand in a submerged line just behind the wave breakline where the backwash is slackening and dropping the sand load. The bar will be submerged for some time as an offshore bar but constant additions of sand cause it to rise above the water surface as a foreshore bar. The distance of such bars from the beach depends on wave size, which governs the point at which waves break. A lagoon of calm water forms behind the bar and fills with alluvium brought down by rivers. The waters of the lagoon increase in freshness and vegetation begins to colonise the shores. Gradually the bar migrates shorewards as waves remove sand from its seaward side to its landward side and eventually a flat marshy coastline bound by dunes or a high exposed bar called a backshore bar results (Fig 235).

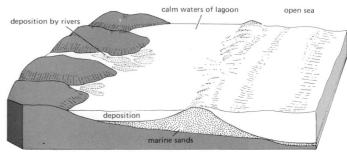

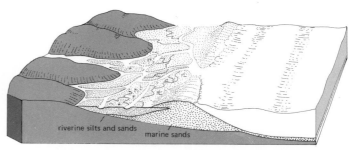

Fig 235.   Block diagrams to show the migration landwards of a sandbar and the eventual filling in of the lagoon. Sand is removed from the seaward side of the bar and deposited on the landward side

Sand-bar and sand-spit coastlines occur along long stretches of the coasts of West and southern Africa. Along the coasts of Senegal and Mauritania longshore movement is southward caused by a northwesterly to northerly swell. The swell is the result of northwesterly winds which are formed by the deflection of the normal northeast trades by onshore breezes. One result is the 43 km-long sand-spit called the Langue de Barbarie which diverts the Senegal mouth southwards (Fig 236). In some years surf caused by southwesterly onshore winds is able to remove some sand from the tip of the spit and so reduce its length. Along the coasts of Sierra Leone and Liberia between Cape Palmas and Sherbro Island longshore drift is, on balance, from southeast to northwest, the prevailing winds coming from the southwest. Offshore bars are a feature of the northern coasts of Sierra Leone, while foreshore bars, sand dunes and diverted river mouths occur southeastwards towards Liberia.

East of Cape Palmas the alignment of the coast again changes and the prevailing southwesterlies create a heavy surf and a strong west-to-east longshore drift. Sand-spits and sand-bars

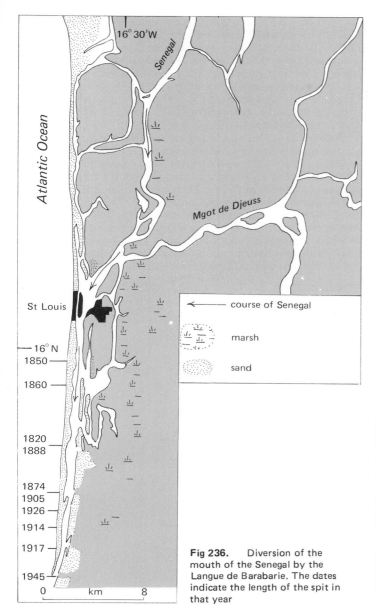

Fig 236. Diversion of the mouth of the Senegal by the Langue de Barabarie. The dates indicate the length of the spit in that year

which now forms a dry sandflat. Bars and spits on this coast are growing fast, up to 1 km a year, and some extend for over 15 km and are up to 1 km wide. Even a large river like the Orange may have its mouth blocked temporarily by a sand-bar during low discharge but at flood the Orange is able to maintain a narrow tidal channel through the bar.

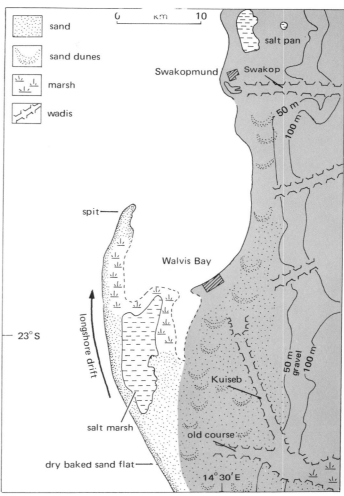

Fig 237. Sandspit formed by longshore drift along the coast near Swakopmund, Namibia

occur all along the shores of the Ivory Coast, Togo and Dahomey and west of Cape Three Points in Ghana. In the Badagri area west of Lagos in Nigeria at least ten successive sand-bars separated by river-laid mud deposits form huge barrier beaches and sand ridges up to 300 m wide, occasionally breached by rivers which drain the lagoons. Drift of sand along the Badagri coastline towards Lagos harbour has been diverted seaward by long sea-walls or moles but the sand now reaches the Niger Delta mouths. Here the Forcados entrance channel has shallowed from 8 m to 3·5 m in 50 years and new sand deposits now exceed 1 million tonnes a month.

Sand-spits caused by northward moving longshore drift (aided partly by the Benguela Current) are also a feature of the Namib and Angolan coasts. Surface winds here are strongly southeasterly or south-southwesterly so that promontories develop northward-trending spits, for example in the Bay of Tigres, at Port Alexandre and at the Port of Lobito in Angola. At Walvis Bay the sand-spit rises about a metre above the water-line and many bays have been enclosed and partially filled in to form small pans (Fig 237). The old course of the Kuiseb River has been prevented from reaching the sea by a sand barrier

Small rivers on the Transkei coast are also blocked during the dry season but the large rivers—the Great Kei, Bashee and Umzimbuvu—are able to maintain tidal channels. Here, and in Natal, northerly longshore drift is caused partly by south to southeasterly winds and partly by a strong tidal current at high water from the south. The bay at Durban has been enclosed by an old uplifted sand-spit called the Bluff and by a zone of coastal dunes called Back Beach to form a large lagoon which is now partially silted up with sediments from the Umlatuzan and Umbilo rivers. The Bluff and a man-made mole divert drifting sands out to sea.

The longshore drift and powerful tidal currents of the coasts of Zululand and Mozambique have formed long spits at St Lucia Bay, at Inhaca enclosing Delagoa Bay, and on Bazaruto Island. St Lucia Lake is a large lagoon separated from the sea by a 2 km-wide line of coastal sand dunes and is drained through a 3 km-wide sand-bar in the south by a narrow tidal channel.

Spits and sand-bars are also a feature on some of Africa's lakes. On the eastern shores of Lake Mobutu in the Bunyoro

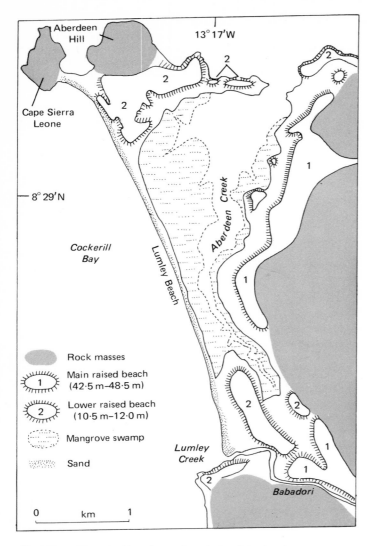

Fig 238. The raised beaches and tombolo of the Lumley area, Sierra Leone

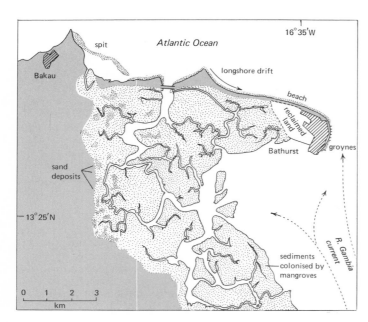

Province of Uganda longshore drift is caused by south to south-westerly winds which have covered a fetch length of up to 150 km by the time they reach Butiaba. Here a 1 km-long spit with a hooked end has been formed and near Kaiso and Tonya lagoons have been created by spits, curved at their ends by strong northerly secondary winds, rejoining the coast. Near Masaka on the west shore of Lake Victoria a long sand-spit has enclosed part of the lake to form Lake Nabugabo and the Lwamunda Swamps.

A tombolo is formed where a spit or sand-bar forms a sand or shingle isthmus linking a former island to the coast. The Lumley sand-bar some 6·5 km west of Freetown in Sierra Leone (Fig 238) has formed a 4 km-long isthmus to link the former rock islands of Cape Sierra Leone, Aberdeen Knoll and the raised beach area of Aberdeen with the mainland. The evolution of the Cape Verde tombolo is shown in Fig 240. The Cape Penin-sula of South Africa can also be considered as a rather more complex tombolo on a very large scale, Cape Peninsula with Table Mountain having once been an island separated from the mainland by narrow faulted straits. Drifting sand gradually

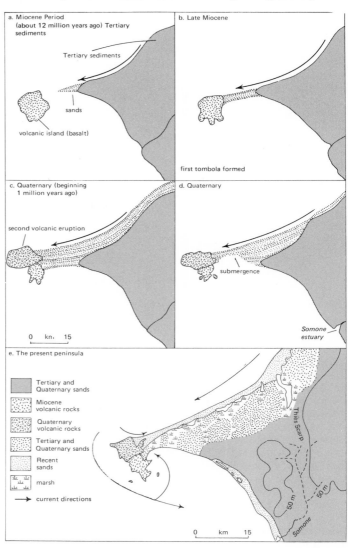

Fig 240. The development of the Cape Verde Peninsula, Senegal

Fig 239. (Left) The hooked spit at Bathurst, the Gambia, West Africa. The river Gambia is steadily filling in the backwaters with riverine sediments

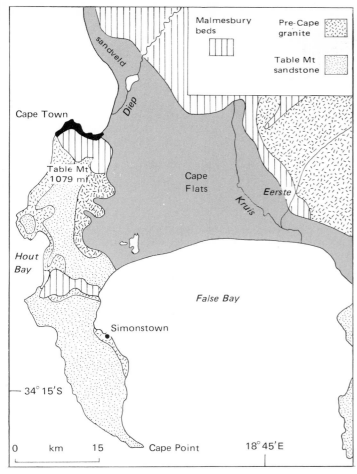

Fig 241. The geology of the Cape Peninsula region, South Africa

level was lowered some 150 m while during the Great Inter-glacial Period melting ice caused the sea-level to rise 30 m. Such changes in sea-level relative to the land by isostatic or eustatic causes results in raised beaches, elevated cliffs and wave cut terraces along the shores of Africa. We cannot expect a regular elevation height, however, since both processes, eustatic and isostatic, may occur at the same time and may also be accompanied by tilting of the continental surface.

Near Sekondi in Ghana, cliffs and beaches stand well above the present sea-level and raised beaches fringe the foot of the Sierra Leone Peninsula mountains (Fig 238). The latter are of reddish-brown clays and silts which contrast with the surrounding greyish sands. The higher raised beaches (42 m to 48 m above sea-level) have occasionally slumped to form steep cliffs 27 m high. A second lower set of raised beaches stands at 12 m above sea-level.

On many parts of the coasts of Kenya and Tanzania raised coral platforms 15 m above the present beaches are backed by low flat zones which were formerly the lagoons of a Pleistocene sea. These coral platforms contain ancient caverns excavated by sea erosion.

The raised beaches along the Namib coast from Conception Bay to Cape Hondeklip south of Port Nolloth contain diamondi-ferous gravel. The terraces occur at various levels, the highest at 100 m. A raised shingle beach between 23 m and 82 m termed the Oyster Line contains ancient shells of oysters which no longer survive in this region. North of the Orange mouth are raised beaches backed by 25 m-high cliffs of a former seashore.

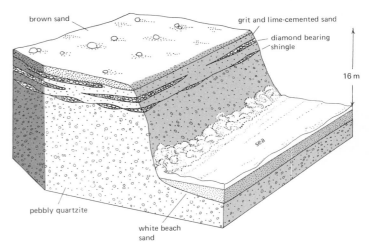

Fig 242. Block diagram of the raised beach at The Cliffs, 18 km north of Port Nolloth and 60 km south of the Orange River mouth, Cape Province, South Africa

settled in the straits to form the extensive sandy Cape Flats (Fig 241).

Where swash on a beach is much stronger than backwash, larger heavier stones pile up along a line of deposition at the up-beach limit of swash action to form a berm, beach ridge or a storm ridge. Such berms become permanent if colonised by vegetation. Large berms are common immediately north of river mouths along the Tanzanian coast, for example, on the Wami delta. Here, however, storm waves are not responsible because the coast is protected by coral reefs. The berms are probably due to large swell waves during high tide and calm weather. Similar berms occur along the east coast of L. Mobutu and are a common feature of the West African sand-bars.

Effects of Isostatic and Eustatic Change

Isostatic uplift is initially caused by the constant deposition of eroded material from the continent on to the continental shelf and the adjustment of the continent to this loss by a slow rising over thousands of years. In Africa the rise was greatest in the south forming the broad coastal plains of Moçambique, and much of the coasts of Liberia, Ghana and the Ivory Coast lie in areas of isostatic uplift. Eustatic change of sea-level is due to fluctuations in the volume of water in the oceans. During the severest glaciations, when water was converted to ice, the water

Depression of a coastline or a rise in sea-level produces a submerged coastline. Submergence may be caused by local faulting, e.g. along the coast of East Africa, by extensive fault-ing, e.g. along the shores of the Red Sea (p. 90), or by the gentle subsidence of a monoclinal depression which may create a relatively straight coastline as seen in the Natal and Angola monoclines (p. 80).

Submergence is, however, usually associated with drowned river valleys and their tributaries (rias), where ridges are convert-ed into promontories and hills into offshore islands. An extensive ria coast extends for some 700 km southwards from Dakar to Cape St Ann near Sherbro Island in Sierra Leone. Here, although elevation of the coast followed a period of downflexing, rias have still been formed due to the depth of the original valleys, the scouring of currents preventing sand drift and accumulation, and the very high tidal range. The estuary of the river Gambia

Fig 243. Uganda: part of the drowned coastline of Lake Victoria

thus has a depth of 79 m in the channel which is over 1·5 km wide and Freetown overlooks the extensively drowned Sierra Leone estuary, the finest natural harbour along this stretch of coast. An examination of the atlas map will reveal many other ria examples along the shores of Guinea and Guinea-Bissau but few good ports because in many cases submerged offshore bars hinder easy access. Guinea possesses only one good port besides Conakry, Benty, which lies 16 km up the drowned Mellacorée estuary. The eastern shores of the Ivory Coast have also been submerged by local downwarping but the drowned creeks and embayments have been filled by drifting sands to form the present smooth coastline.

Local faulting and depression along the East African coast has produced several excellent, although isolated (for this is generally a coast of uplift), examples of rias. Rivers here have cut narrow channels through the coastal raised coral platform and the sea has penetrated inland to drown the dry sites of former lagoons. The deep channel at Kilindini and the shallower Mombasa Harbour channel are cut through the raised coral platform and Ports Reitz and Tudor are the drowned mouths of the rivers Mwachi, Cha Simba and Marea. Other rias due to local sinking occur at Mtwapa, Kilifi, in the Lamu archipelago, (Fig 244), at Dar es Salaam, along the coasts of Pemba Island, and in the Kilwa archipelago in southern Tanzania. The coast of East Africa, however, is not really a true ria coast for these are isolated examples rather than typical features in an area of uplift.

The downwarping of the central plateau of East Africa and the formation of Lake Victoria (p. 68) has also produced ria coastlines and offshore islands (Fig 245). Kavirondo Gulf is a submerged faulted trough, Smith Sound west of Mwanza in Tanzania is an 80 km-long ria, and the Sese Islands of Uganda were once hills, as were Ukara and Ukerewe islands in the southeast region of the lake. The 'grain' of the land, that is, the trend of hilly granitic ridges and valleys is discordant—it runs at an angle to the coast. On the opposite southwest side of the lake, south of Bukoba, the trend of the Pre-Cambrian sandstones is from southwest to northeast, a concordant coastline with few inlets. Three folded ridges here rise to over 1500 m and the most easterly ridge has almost been completely submerged to form the linear Bumbiri Islands some 10 km offshore.

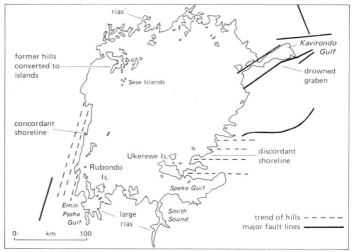

Fig 245. The submerged shorelines around Lake Victoria

Coral Reef Coasts

Coral reef and lagoon coasts occur along the shores of Moçambique, Tanzania, Kenya, Somalia and the Red Sea. Some reef has also developed along the coasts of northern Natal and Zululand but it is poorly developed and discontinuous. Coral has also been dredged from the Agulhas Bank south of latitude 35°S off the South African coast indicating former warmer waters in this region.

The natural conditions of the Kenyan and Tanzanian coasts

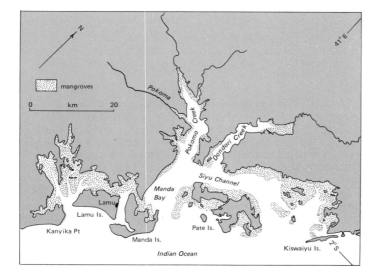

Fig 244. The Lamu archipelago, an example of a locally drowned ria on the north Kenya coast

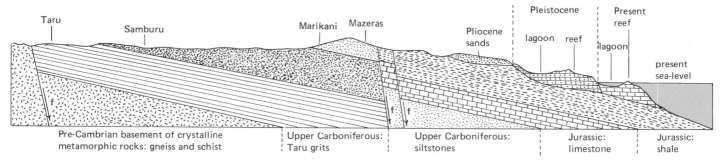

Fig 246. Cross-section extending from a point 10 km north of Mombasa, Kenya, inland for some 80 km in a north-west direction. Study the diagram and attempt a short account of the geological history of this region.

are ideal for the growth of immense numbers of polyps, minute sea creatures living in colonies. Polyps create their skeletons by extracting calcium carbonate from sea water and build up considerable reefs of limestone, the growth rate being between 1 and 10 cm a year. Other marine organisms such as algae and formanifera aid in the reef-building process. The polyps flourish in shallow (25 m–60 m), clear salt water with an average temperature of approximately 20°C–35°C and they are limited to within 30° north and south of the equator. Thus the sandy waters off the coast of West Africa and the cold waters of the southwestern coasts of Africa are not suitable for coral development.

Along the East African coast coral platforms extend out to sea for distances ranging from 500 m to approximately 2 km. At the edge of the platform, away from the sandy shores, conditions are ideal for coral growth and a barrier reef builds up (Fig 246). The reefs enclose shallow lagoons and, at low tide when the coral platform is exposed, it is possible to walk over it. The landward edge of the lagoons is fringed by dazzling white beaches composed of fine abraded coral particles. Islands of uplifted older Pleistocene coral lie near the shore, their bases notched and undercut by mechanical and chemical erosion. Inland from the beaches, low raised cliffs mark the line of earlier shores. Wherever streams empty their muddy waters into the sea and reduce salinity the coral cannot exist and this causes breaks in the reef.

Fieldwork

1. Using a calibrated pole and a stop watch, measure the height of the waves and time their speed. Relate your observations to wind direction, force, angle of beach, etc.
2. Examine coastal rocks for signs of notching, pitting and other forms of erosion. What is the angle of rock beds to the horizontal? Has this had an effect on the cliff profile? Measure the height, length and width of arches, stacks, caves, and the depth of rock pools.
3. Make a beach transect by measuring the length of the beach from sea edge inland at low tide and measuring each section of the beach length which has different sized particles, e.g. a silt zone, a sand zone, a pebble zone, etc. Are the sections graded in size from sea inland? Draw a sketch of your transect and indicate on it the average size of the particles in each section. Use a pebbelometer (a ruler with a fixed end and a wooden slide moving up and down) to measure the bigger stones. Is there evidence of storm action (berm formation)? Can you account for this? Are particles rounded or angular?

Questions

1. Describe the action of waves.

2. With examples from Africa, describe the formations of a wave-cut platform and cliffs. In what parts of Africa is differential erosion active?

3. With reference to examples along the coast of Africa and with the aid of sketch-maps and diagrams, describe the mode of formation of the following depositional features:

    sand-bars; sand-spits; tombolos; berms.

4. What have been the effects of isostatic and eustatic changes of sea-level on the coasts of Africa?

5. Explain the following terms, referring to examples from the coastlines of Africa:

    concordant coastlines; discordant coastlines; ria coastlines.

6. Explain the formation of coral-reef coastlines with reference to the East African coast.

16   Man-made Landscapes

Over the last 2500 years, using his own energy and that of water and animals, man has acted as a geomorphological agent, creating ever larger artificial landforms and remoulding the landscape. During the last three-quarters of a century, a mere drop in the ocean of geological time, the rapid advance of man's technology has enabled him to rival the natural forms of erosion and deposition and to produce his own landforms in every type of climatic environment, altering the existing patterns of natural geomorphology. Growing concern for the environment has focussed man's attention on his own activities in the landscape and a new branch of landscape study has evolved: anthropogeomorphology—the study of processes affecting landforms in landscapes where natural processes have been halted or disturbed by man.

Man first acted as a geomorphological agent when he shaped his first hand-axe. The southern Rift Valley region of Kenya is particularly rich in stone hand-axes, especially in the Pleistocene lake beds at Olorgesaillie where hundreds of stone axes litter the ground like the traction load of a dry river bed. As an animal hunter from earliest times, man began to alter the delicate balance existing in the animal kingdom, effecting the vegetation cover and, indirectly, the weathering and erosion processes. By his use of fire, man has profoundly altered the vegetation cover of Africa, changing forest to woodland, woodland to savanna, savanna to grassland, and grassland to semidesert, with consequent effects on geomorphological processes. Bush fires may completely alter the soil profile, reducing bacterial activity and freeing large quantities of silica which are easily swept away by wind. The widespread removal of tropical rainforest exposes lateritic soils to rainwash and baking which has hastened the development of lateritic cuirasses. During Roman times the increased felling of the trees in the Maghreb region depleted the forests so severely that today many parts are still barren and eroded badlands.

Fig 247.   Mauritania: the Tazadit Mine of Miferma Mining Company. This, and nearby similar opencast mines, produce 9 million tonnes of iron ore a year. A man-made landscape of artificial terraces

Fig 248. Zimbabwe—Zambia border: the dam wall and part of Lake Kariba — one example of man's impact on the fluvial cycle in Africa

Removal of forest cover also results in basic changes in the water table. Run-off rates are increased at the expense of seepage and overland flow to rivers is increased to cause flooding. The water-table may thus be shifted considerably to new areas. Migrations of people from Kano to Bornu Province in northern Nigeria caused the water-table to rise by 60 m in 50 years. Percolation here was reduced by forest removal and run-off to rivers increased. Seepage through the stream bed increased the water-table in the valleys while reducing it in the interfluves, a reversal of the normal trend.

As a cultivator and herdsman, man has had a lasting effect on the surface morphology of Africa. Overgrazing by sheep, goats and cattle in the Atlas region has reduced former woodland areas to sparse shrub cover, increasing run-off and encouraging creep and slump. The concentration of domestic animals at water holes and cattle dips compacts the soil and forms ready-made channels which develop as dongas—a network of deeply eroded, steep-walled gulleys which are constantly widened by the crumbling of their sides. Overgrazing, the tearing out of grass roots in time of drought by goats and sheep, and the breaking down of the soil structure by sharp hooves accelerates wind erosion.

Careless cultivation techniques destroy soils and lay bare the regolith to further removal. The growing of one crop such as maize on the same land year after year gradually removes the nitrogen content, and the wide spacing of plant rows instead of intercropping prevents natural routes for rill and gulley formation. The export of commercial crops also exports soil minerals, for example, the phosphorus and nitrogen in exports of maize, cotton and groundnuts. Between 1966 and 1973, Nigeria lost 168 000 tonnes of phosphates in her exports of 3 million tonnes of groundnuts. In Senegal, extension of the area under export groundnuts has changed farmland to sandy wastes.

The needs of industry also create artificial landscapes. By mineral extraction from the earth's crust using opencast mining techniques—(the removal of surface beds to enable extraction of ores)—the landscape is scarred by huge pits displaying convex or concave slopes and intricate patterns of terraced roadways (Fig 247). The amounts removed annually and never to be replaced are huge: 2·5 million tonnes of iron ore from Sierra Leone's mines and 20 million tonnes from Liberia, 700 000 tonnes of copper ore from Zambia and 385 000 tonnes from Katanga. Flooding of such pits creates artificial lakes, for example, in the Nkandabwe open-pit coal mine in Zambia or the Kimberley Diamond Mine's Big Hole.

Sub-surface mining operations also produce their particular landforms. In the Carltonville-West Driefontein goldfields of South Africa, lowering of the water-table by pumping has caused the dolomitic limestone beds to become dehydrated and more compact. Water trickling through avens (percolation channels) in the limestone has weakened the dehydrated rock causing collapse, considerable surface damage to property, and producing over one hundred 'pseudo-karst' sinkholes. Dynamite blasting in the Witwatersrand mines produces artificial shock tremors which also cause minor surface damage. Waste material from the mines is dumped in huge spoil heaps creating a landscape dominated by mesa-like forms. Wet-weather slumps and dry-weather sandstorms from these dumps prompted the authorities to anchor the slope surfaces with tough, reed-like grasses.

Man has also influenced the geomorphological processes related to marine and fluvial activity. Continuous dredging of

Fig 249. South Africa: man's concrete jungle begins to dominate the natural landscape: tarmac highways, concrete flyovers and high-rise flats at the foot of Table Mountain, Cape Town.

port channels and the construction of moles arrests the build-up of silt deposits and diverts tremendous quantities of sediments (p. 51). The cutting of the Vridi Canal from the Ebrié Lagoon through a huge sand-bar to the Gulf of Guinea at Abidjan in the Ivory Coast has diverted millions of tonnes of silt into the Trou Sans Fond (trough without depth), a deep submarine trench off shore. Silt diversion has also occurred along the moles at Tema and Takoradi in Ghana. In some cases man has extended the land surface at the expense of the sea, for example, the 144 ha reclaimed foreshore at Cape Town and the reclaimed areas at Bathurst (Fig 239).

MAJOR MAN-MADE LAKES AND RESERVOIRS OF AFRICA

Lake name	Maximum area (km²)	Maximum depth (metres)	Ratio of annual inflow to volume	Maximum drawdown[1] (metres)	Date of construction
Kariba	4300	125	1:4	3	1958
Volta	8800	80	1:4	3	1964
Nasser	6000	80–90	1:2	—	1964
Kainji	1250	60	4:1	10	1968
Kafue Gorge	3100	58	2:1	7	under construction
Cabora Bassa	2700	157	—	—	—
Seasonal Nile Reservoirs					
Gebel Aulia	600	12	8:1	6	1937
Sennar	140	16	50:1	17	1925
Roseires	290	50	16:1	13	1966

[1] Maximum fall of lake surface to highest sluice gate

Based on a Table in *The Inland Waters of Tropical Africa*
L. C. Beadle, Longman 1974

The construction of dams across river valleys and the establishment of extensive irrigation networks alters the existing fluvial cycle of erosion and deposition. A dam creates a new artificial base-level above which deposition increases and below which water flow and erosion is greatly reduced. The system of diversion channels leading to artificial basins along the Nile has diverted the Nile sediments from the delta and the Mediterranean. The Volta lakes created by the damming of the river at Akosombo now covers three per cent of Ghana and the lake's annual rise and fall creates coastal mud flats stretching for some 7200 km. The supply of sediment to the Volta delta (p. 53) has been considerably reduced thus restricting the growth of a natural feature. The completion of the dams on the middle Orange in South Africa will irrigate a further 280 000 ha of land, reducing soil erosion by wind and raising the water-table. At Kariba when the sluice gates are opened the force of millions of tonnes of water has created an artificial plunge pool and has caused structural changes in the rock bed below the dam wall. The man-made lakes of Africa (Table, left) have created new landscapes of ria coasts, alluvial shorelines and deltas as well as affecting local climates by the increased evaporation of surface water.

The impact of man on Africa's landscape seems small when compared with the industrialised regions of Europe and North America and the continent's surface still retains a natural or semi-natural aspect. But man's ability to create larger and more extensive artificial landforms and landscapes is likely to increase rapidly as his technological knowledge and his numbers expand. The study of anthropogeomorphology will involve challenging work for future geomorphologists living and working in Africa.

Questions

1. In what way has man altered the landscape in your country?

2. What changes have been brought about in your region by

(a) burning of the vegetation
(b) clearance of the vegetation by cutting
(c) conservation of water resources?

3. What conservation projects are being carried out in your region to

(a) protect land from erosion
(b) conserve the soil
(c) protect the vegetation cover?

17   Map Reading Section

This section contains the following map extracts:

1. South Africa: Graskop sheet 1:50 000 Karst landforms.
2. Tanzania: Moshi sheet 1:50 000 Young streams in a volcanic area.
3. Nigeria: Toro sheet 1:50 000 Scarp retreat, pediplanation, inselbergs.
4. Malaŵi: Zomba sheet 1:50 000 Pediplanation, inselbergs, highland dissection.
5. Uganda: Kampala sheet 1:50 000 Accordant summits.
6. Zimbabwe: Pungwe Falls sheet 1:63 360 Erosional cycle and river capture.
7. Kenya: Mt Kenya sheet 1:25 000 Glaciated upland.
8. Tanzania: Arusha-Tengeru sheets 1:50 000 Volcanic region.
9. Ghana: Wurupong sheet 1:50 000 Drainage patterns and adjustment to structure.
10. South Africa: Oudtshoorn sheet 1:500 000 Folded mountains and superimposed drainage.
11. Tanzania: Ngombezi sheet 1:50 000 Dissected horst, river adjustment to structure.
12. Tanzania: Mazinde sheet 1:50 000 Fault block, graben and inland delta.
13. Kenya: Nairobi-Nakuru sheets 1:250 000 Rift faulting and vulcanicity.
14. Zimbabwe: Nkai sheet 1:50 000 Pan formation and seepage in a dry savanna region.
15. Kenya: Kilifi sheet 1:50 000 Coral reef and lagoon coast.
16. Uganda: Sempaya sheet 1:50 000 Final test map.

Analysing Relief Maps

Use the following guide when analysing each of the relief maps in this section. This analysis should be done for each map in addition to the questions set.

1. Lay tracing paper over the map extract and shade in summit levels. Analyse your findings. Are the summits accordant? Do they show evidence of planation?

2. Examine the elevation of the bottoms of valleys. Is accordance also shown here? What are the maximum and minimum elevations? How steep are the slopes? What are the heights from bottom to top of the valley sides? Is there a uniformity of valley slope and height throughout the region or are there significant differences?

3. Construct a slope map (p. 44) and analyse the map area by degree of slope.

4. Construct transects noting the chief morphological features and superimpose them on each other to indicate general aspects of the landscape and erosion-level accordance if any.

5. By studying the surface landforms, degree of river incision, etc., try to visualise the rock types and the geological structure. Is the escarpment a tectonic or an erosion type? Is the drainage pattern due to fault alignment or to the trend of softer rock zones? If possible, try to obtain a geological map of the area *after* you have made your assessment.

6. Finally, draw a sketch-map of the region dividing it into major relief zones, for example, plateau, area of dissected upland, lowland plain, etc. Indicate on this map smaller relief features such as an alluvial fan, a knife-edged ridge, an outlier, etc.

If it is possible to visit the area of relief shown on the map, note the occurrence of landforms not normally shown because of their small size, for example, debris fans, debris avalanches, small inselbergs, pebble strewn slope, etc., and mark these features on your own geomorphological map.

1. South Africa: Graskop Sheet 1:50 000

Tower karst landforms

Read pages 32–5 before starting this exercise.

1. Locate the area in your atlas by referring to the latitude and longitude shown.

2. Study the drainage pattern of the region. In what manner are the rivers dissecting the landscape? Explain why the Blyde River is a permanent stream while most of its smaller tributaries are intermittent (shown by broken blue lines).

3. Lay a piece of tracing paper over the map and mark on it the summits of karst cones to plot their distribution. Note any similarities or dissimilarities in altitude. Are there any reasons for these? Study the cross-section on page 61 for structural clues.

4. On your tracing locate and name the following landforms: a knife-edge ridge; a flat-topped ridge; cols or saddles; defile or pass; a prominent spur; a dissected escarpment.

5. Draw a cross-section from the cone at 1528 metres north of the Lisbon River through the village of Frankfort to the western edge of the map. Your transect line should be in two sections, one passing through the line of cones to the valley of the Blyde, then altering direction slightly to pass through Frankfort. Describe the features shown and insert the probable rock strata (see Fig 53).

6. Write a brief description of the development of the landforms in this region. You may refer to pages 32–5 but you should write the account in your own words.

2. Tanzania: Moshi Sheet 1:50 000

Young streams in a volcanic area

Read the following pages before starting this exercise: 48, 54, 98–103.

1. Locate the area in your atlas by referring to the latitude and longitude shown.

2. Find the lowest and highest points on the extract and work out the average gradient between the two.

3. Draw a transect from the church at Fukeni (grid square 2136) to Nanga (265 327). Describe the valleys on the transect and work out the gradients of the slopes of the valleys of the Msaranga and the major river south of Kimochi. What reasons are there for the deeper incision of these two rivers?

4. Work out the long profile gradient of the Msaranga River from its source to where it ends in square 2029. Draw a section and account for any irregularities of the bed.

5. Trace a sketch-map to show only the rivers in blue. Taking the river points at 243, 247, and 268 along the top edge of the map as first order streams, divide your river sketch-map into first order, second order, etc., streams. Is there a possibility of future river capture on the map? Mark this.

6. Write a general description of the drainage pattern noting valley depths and the landforms directly attributable to stream erosion. Estimate the probable speeds of the rivers by comparing with rivers in your own area which have similar gradients. What type of rock are these rivers eroding?

7. What landform is seen at Nanga (265 327)? Imagine you are standing at the triangulation point 4747 at Nanga looking northwestwards. Make a sketch of the landscape you would see.

8. What do you think are the effects of the forested region indicated by tree symbols in the north of the map, on the drainage of the area?

3. Nigeria: Toro N.W. Sheet 1:50 000

Scarp retreat, pediplanation and inselberg formation

Read pages 29–31, 37–45 before you start this exercise.

1. Locate the area in your atlas by referring to the latitude and longitude shown.

2. Draw a sketch-map to show the chief physical divisions of this region and briefly describe the characteristics of each of your regions.

3. Draw two transects, one running north–south and the other running east–west, which would best show the processes of pediplanation, scarp retreat, and inselberg formation in the region.

4. Write a short account of the theory of scarp retreat and pediplanation with special reference to this region.

5. Select one or two of the inselbergs shown and draw enlarged sketch-maps of them like those in Fig 39.

6. Imagine you are standing on the summit of the large inselberg in the extreme north of the map extract (at over 3000 feet) and that you are looking towards the Duchin Solli Hills. Draw a simple field sketch of what you would see and write a brief description of the scene.

7. Describe the drainage pattern of the region and the way in which it is eroding the landscape.

4. Malawi: Zomba Sheet 1:50 000

Pediplanation, inselbergs and highland dissection

Read pages 29–31, 37–45 and study Fig 67 before starting this exercise.

1. Locate the area in your atlas by referring to the latitude and longitude shown.

2. There is evidence of pediplanation at several levels on this map extract. Produce a tracing with the levels shaded in two different colours. Read pp. 40–44 and state what cycles of erosion these represent. What reasons can you give for the much lower height of isolated hills such as Sanjika (523 935)?

3. Draw transects through Sasi Hill (488 948), Ntanya (485 929), spot height 3909 (478 905) to point 475 890 and also north–south through Nkhoronje (508 944) and Ulumba (508 897). Comment on your findings in the light of what you have read on the accordance of summit and valley-bottom levels.

4. Using your transects, imagine you are at point 530 920 looking westwards through an angle of approximately 140°. Sketch the landscape you would expect to see adding notes to your diagram.

5. Work out slope angles at various points on the extract and compare these with known slopes in your own area.

(continued on p. 145)

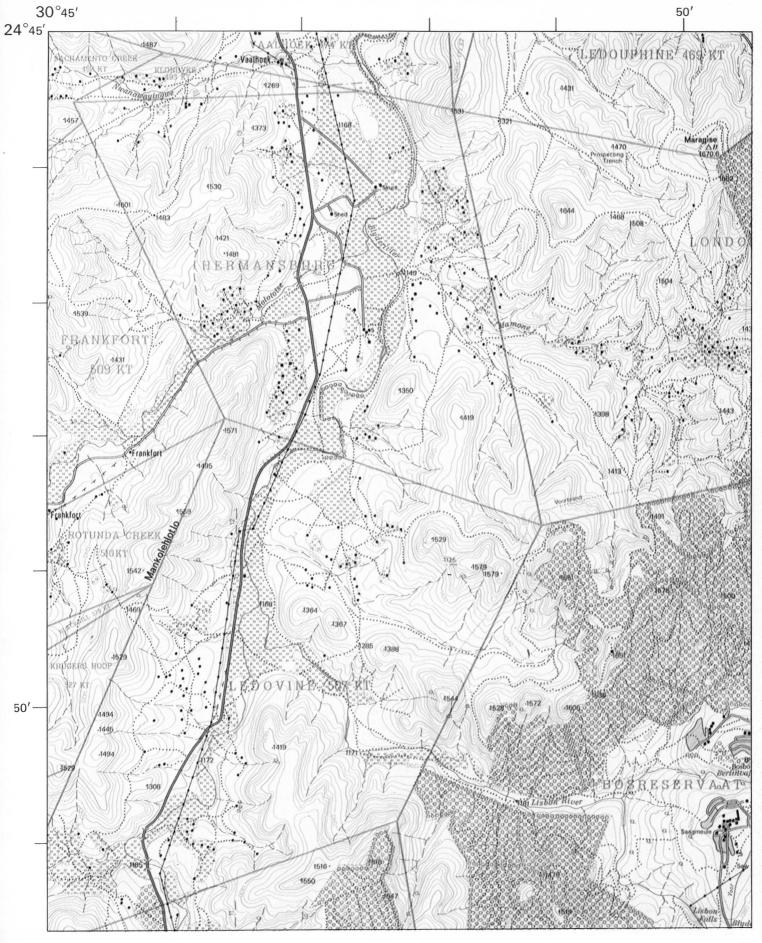

1. South Africa: Graskop Sheet   Scale: 1:50 000   Contour interval 20 ft

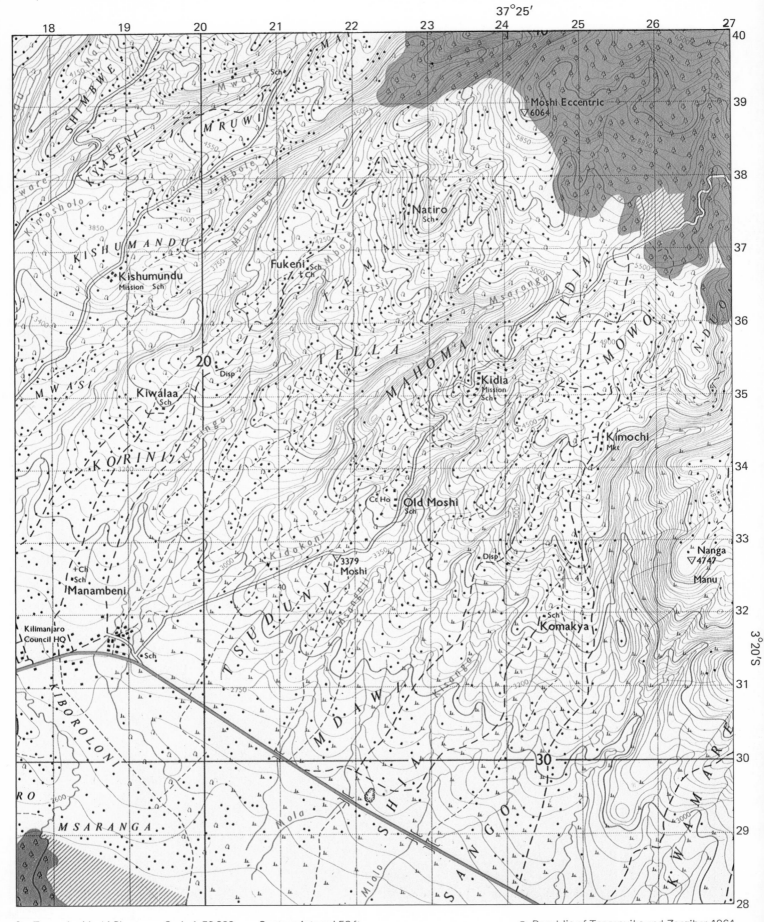

2. Tanzania: Moshi Sheet   Scale 1:50 000   Contour interval 50 ft   © Republic of Tanganyika and Zanzibar 1964

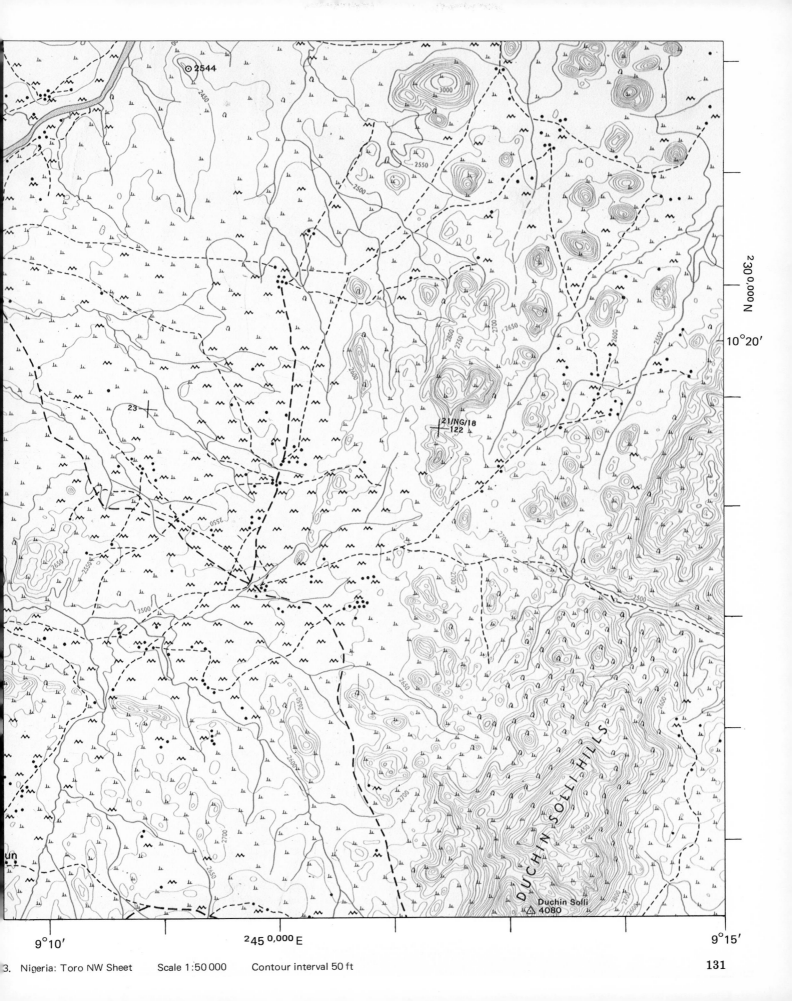

⊙ 2544

230 0,000 N

10°20'

23

21/NG/18
122

2550

2500

2550

2500

2550

2650

2700

2750

2800

2650

2600

2550

2700

2500

2650

2700

2700

2650

2550

2600

2500

DUCHIN SOLLI HILLS

Duchin Solli
△ 4080

un

9°10'   245 0,000 E   9°15'

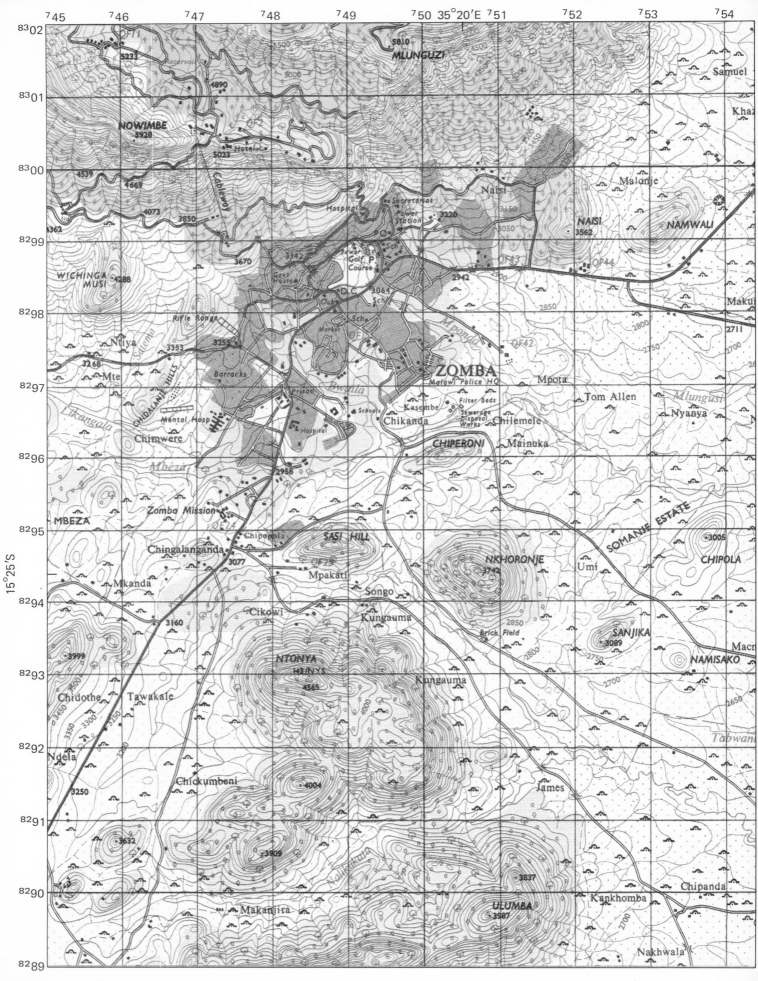

4. Malawi: Zomba Sheet   Scale 1:50 000   Contour interval 50 ft   © Malawi Government 1967

5. Uganda: Kampala Sheet   Scale 1:50 000   Contour interval 50 ft

Reproduced from Directorate of Overseas Surveys map DOS 426 Sheet 71/1 second edition by permission of the Controller of HM Stationery Office

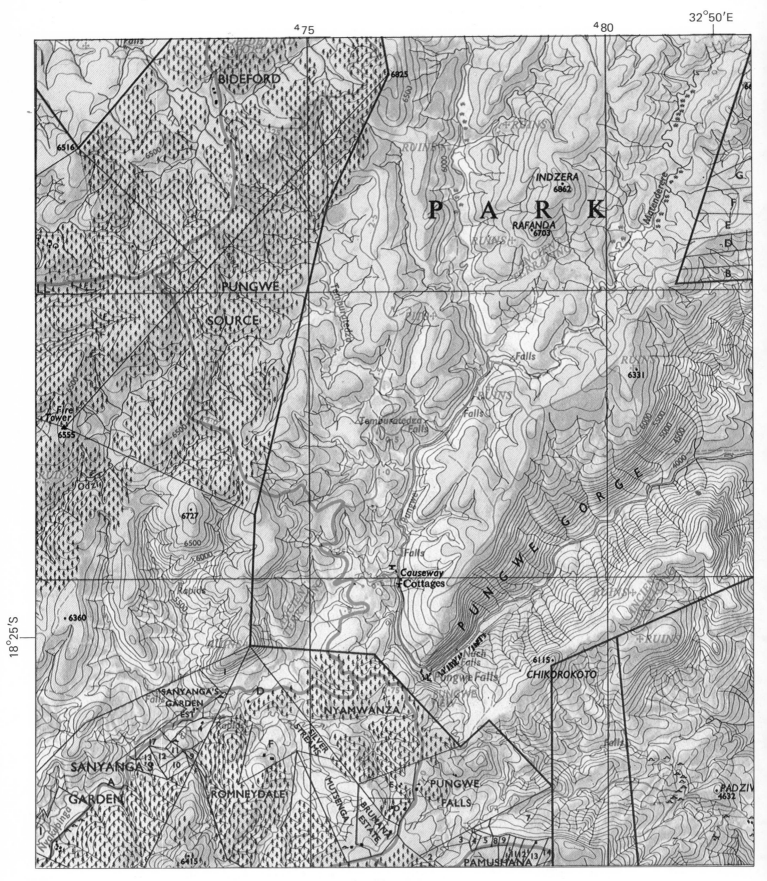

6. Zimbabwe: Pungwe Falls   Scale 1:63 360   Contour interval 100 ft

134

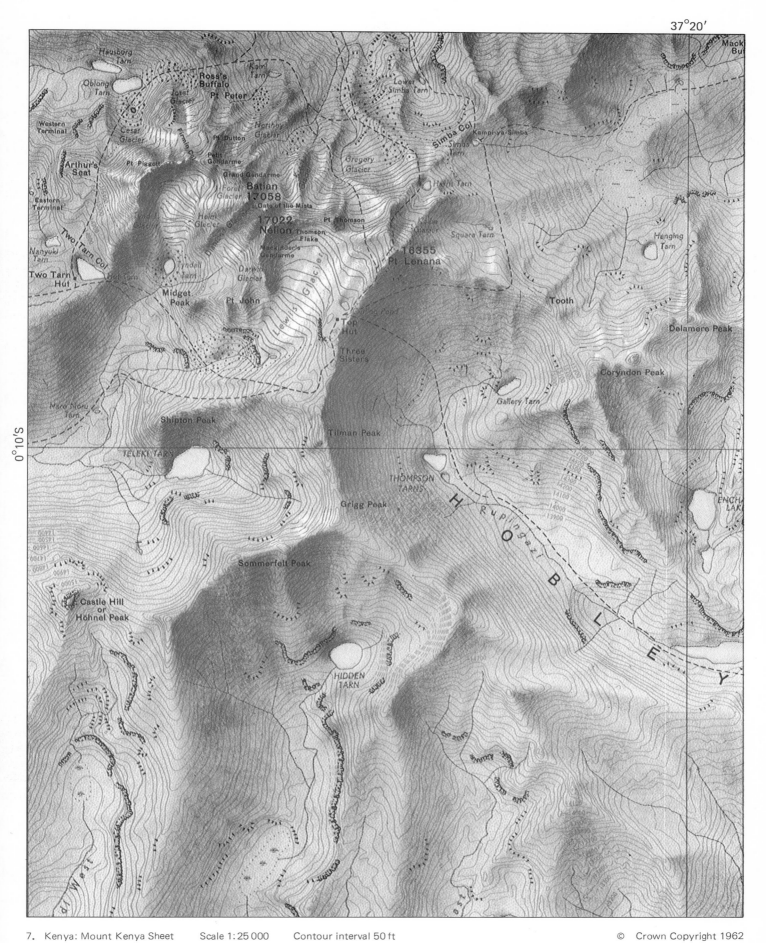

Hausburg Tarn
Oblong Tarn
Ross's Buffalo
Kami Tarn
Lower Simba Tarn
Pt Peter
Josel Glacier
Western Terminal
Cesar Glacier
Pt Dutton
Northey Glacier
Simba Col
Kampi-ya-Simba
Simba Tarn
Arthur's Seat
Petit Gendarme
Gregory Glacier
Grand Gendarme
Batian 17058
Forel Glacier
Gate of the Mists
Helm Tarn
Eastern Terminal
17022 Nelion
Pt Thomson
Square Tarn
Hanging Tarn
Nanyuki Tarn
Two Tarn Col
Thomson Flake
16355 Pt Lenana
Two Tarn Hut
Mackinder's Gendarme
Tyndall Tarn
Darwin Glacier
Midget Peak
Pt John
Tooth
Delamere Peak
Shipton Peak
Tilman Peak
Corydon Peak
Gallery Tarn
Top Hut
Three Sisters
Mara Moru Tarn

TELEKI TARN
THOMPSON TARNS
14300
14200
14100
14000
13900
H R uplingazi O B L E Y
Grigg Peak
ENCH LAK
Sommerfelt Peak
14700
12600
12700
Castle Hill or Höhnel Peak
14900
15000
HIDDEN TARN

7. Kenya: Mount Kenya Sheet   Scale 1:25 000   Contour interval 50 ft   © Crown Copyright 1962

Reproduced from Directorate of Overseas Surveys map DOS 302 Sheet Mount Kenya by permission of the Controller of HM Stationery Office   **135**

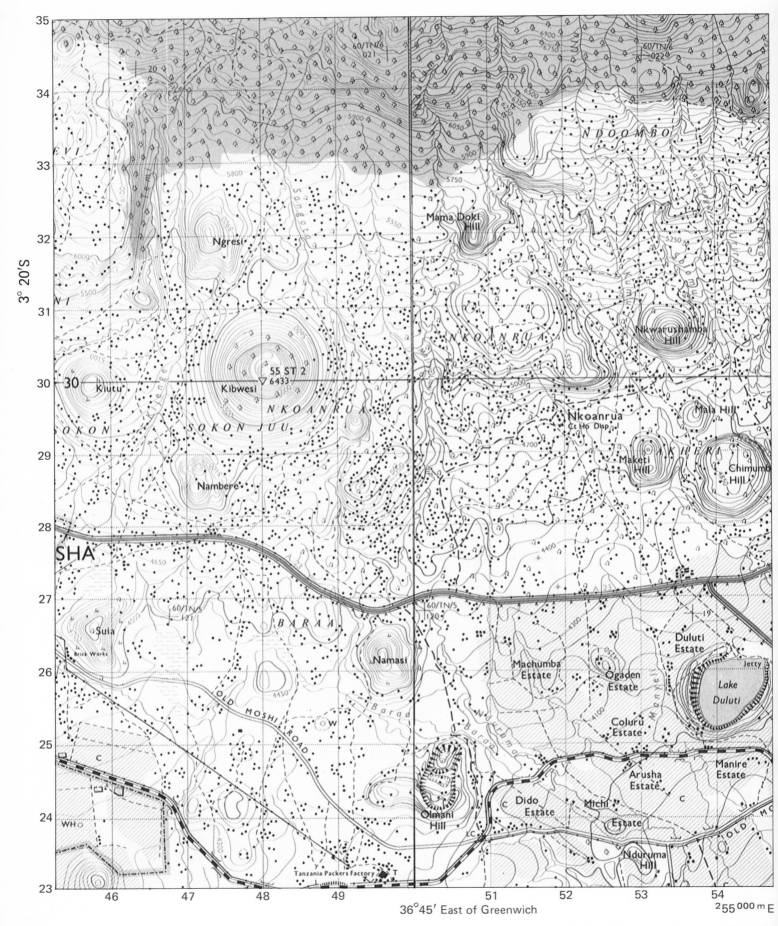

8. Tanzania: Arusha-Tengeru Sheets   Scale 1:50 000   Contour interval 50 ft   © Government of the United Republic of Tanzania 1969, 1967

Reproduced from E Africa DOS 422 Tanzania Sheet 55/3 Ed 1 and DOS 422 Tanzania Sheet 55/4 Ed 1

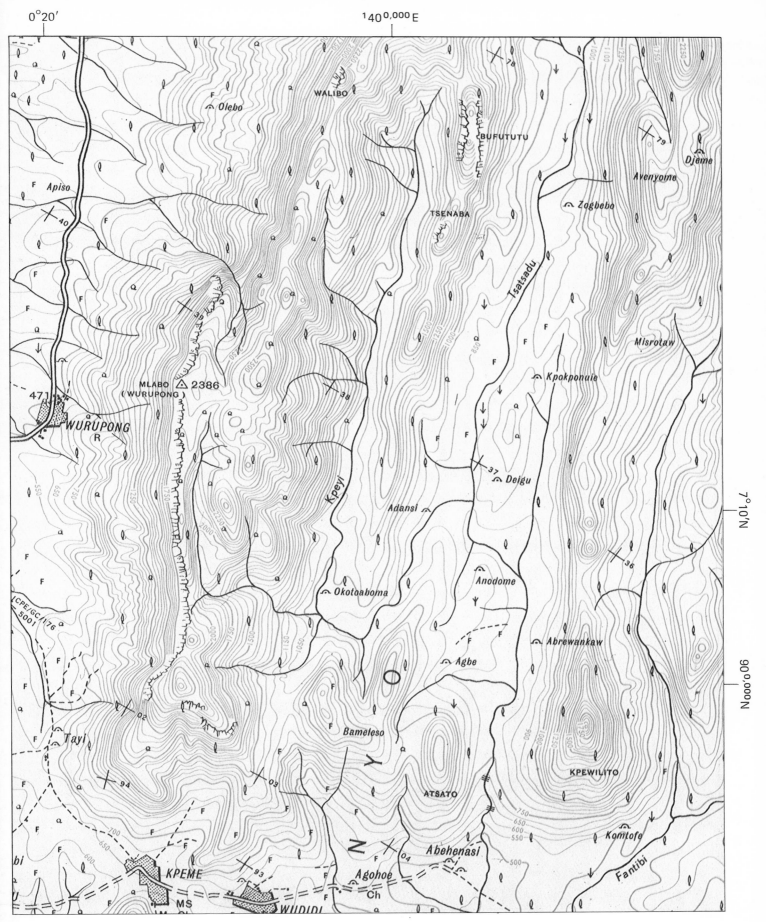

9. Ghana: Wurupong Sheet 182   Scale 1:50 000   Contour interval 50 ft   Crown Copyright

Reproduced from Directorate of Overseas Surveys map DCS 14 Sheet 182 by permission of the Controller of HM Stationery Office

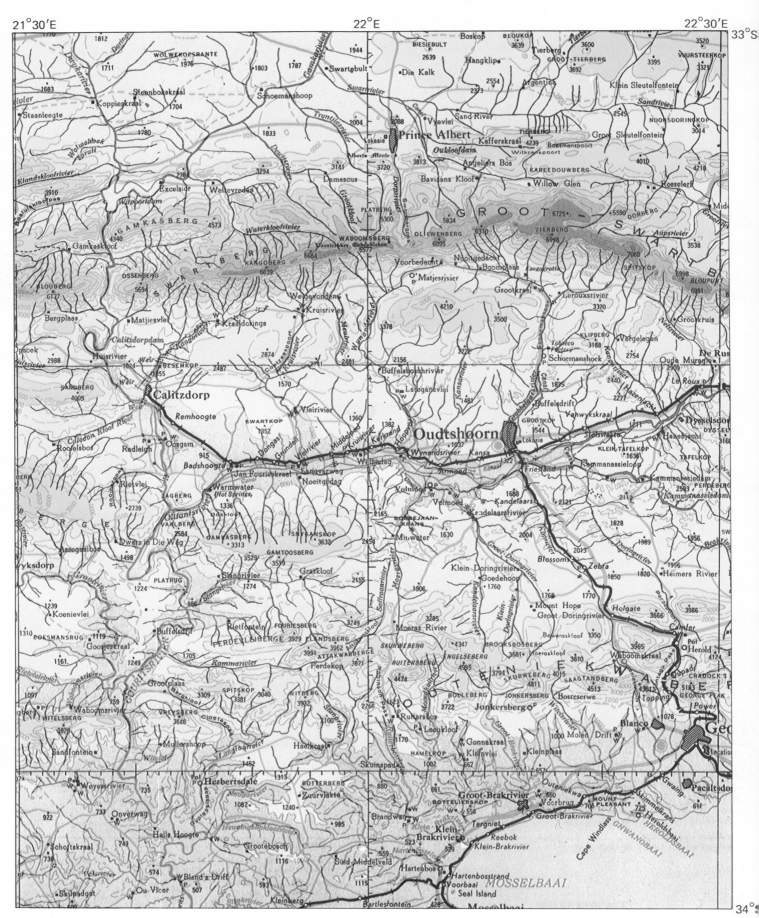

10. South Africa: Oudtshoorn Sheet   Scale 1:500 000   Contour interval 500 ft

138

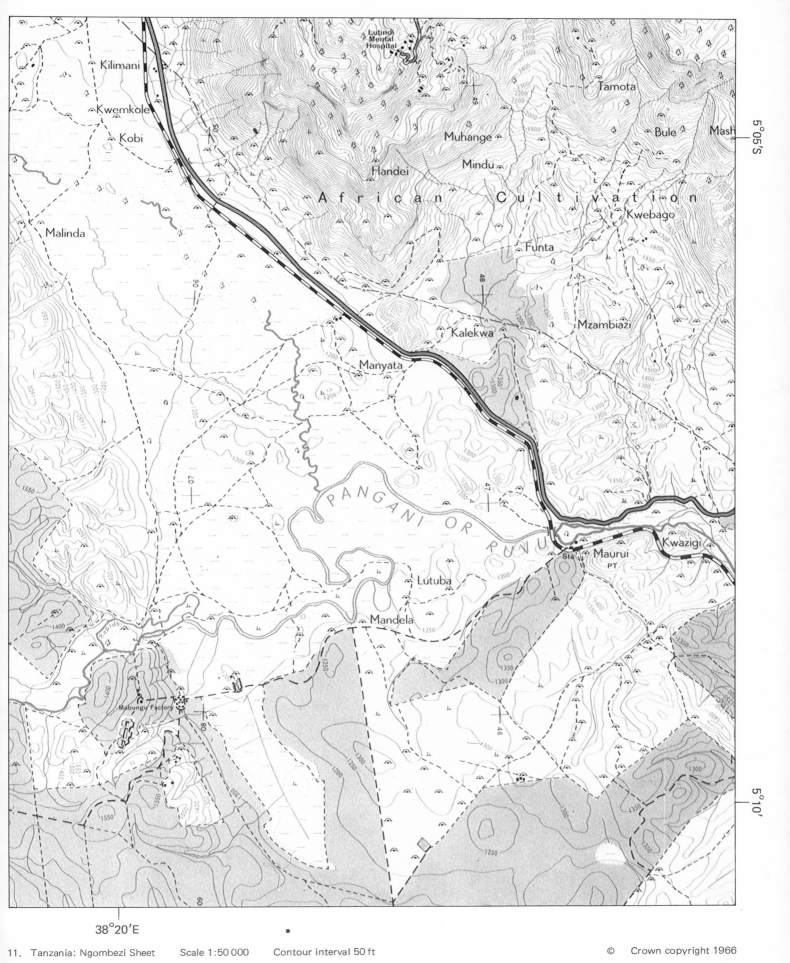

11. Tanzania: Ngombezi Sheet   Scale 1:50 000   Contour interval 50 ft   © Crown copyright 1966

Reproduced from Directorate of Overseas Surveys map DCS 22 Sheet 129/2 First edition by permission of the Controller of HM Stationery Office

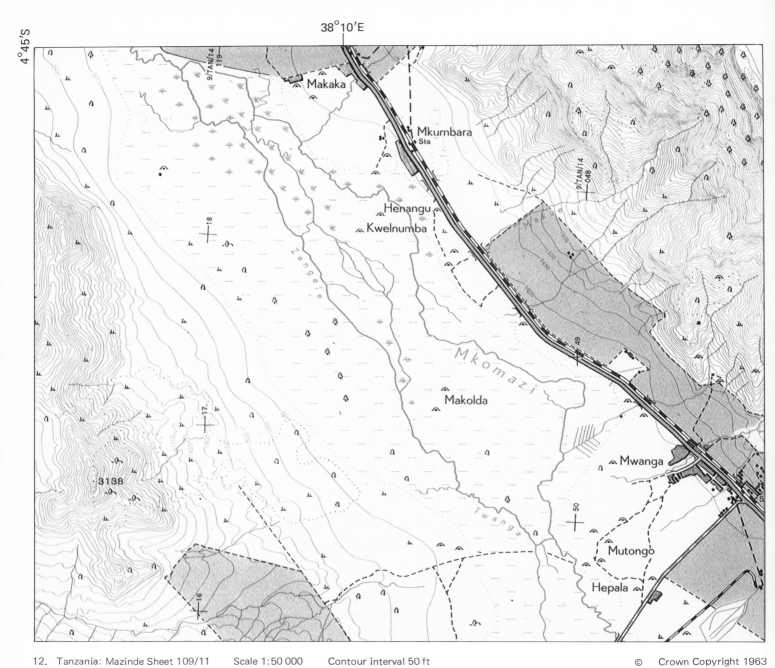

12. Tanzania: Mazinde Sheet 109/11   Scale 1:50 000   Contour interval 50 ft   © Crown Copyright 1963

Reproduced from Directorate of Overseas Surveys map DCS 22 Sheet 109/111 Edn 1 by permission of the Controller of HM Stationery Office

13. Kenya: Nairobi-Nakuru Sheets   Scale 1:250 000   Contour interval 200 ft   ©   Kenya Government 1974

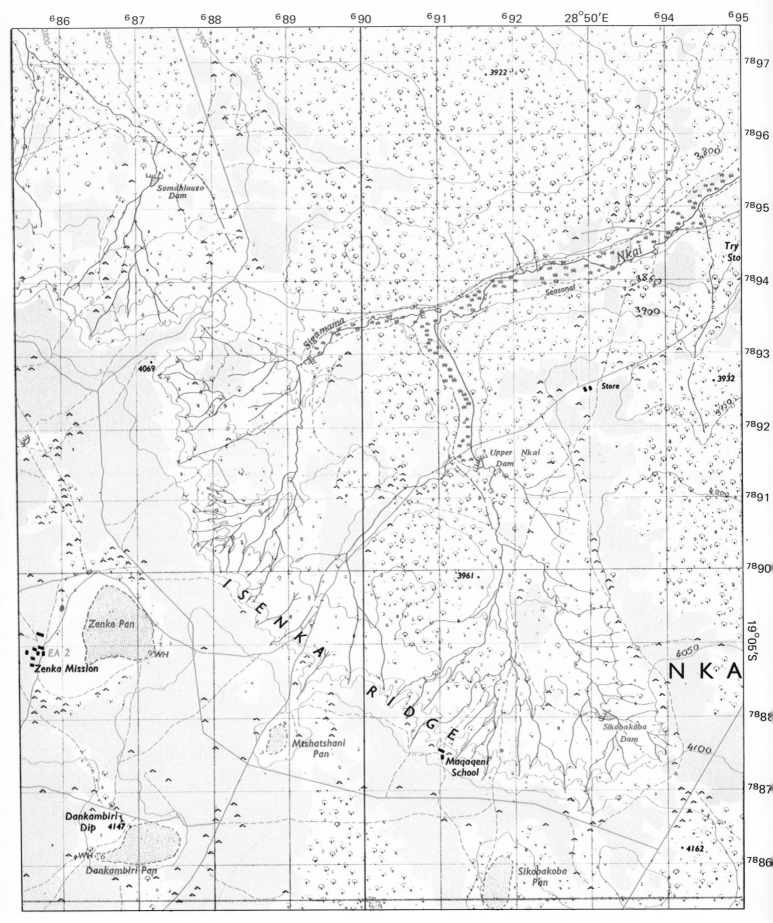

14. Zimbabwe: Nkai Sheet   Scale 1:50 000   Contour interval 50 ft

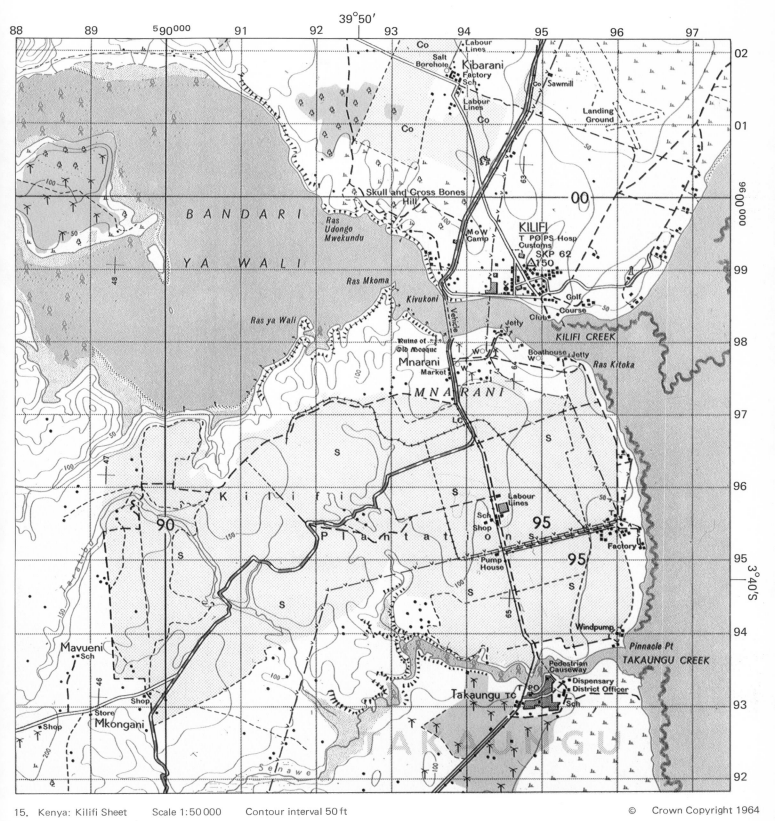

15. Kenya: Kilifi Sheet   Scale 1:50 000   Contour interval 50 ft   © Crown Copyright 1964

Reproduced from Directorate of Overseas Surveys map DOS 423 Sheet 198/2 Edn 3 by permission of the Controller of HM Stationery Office

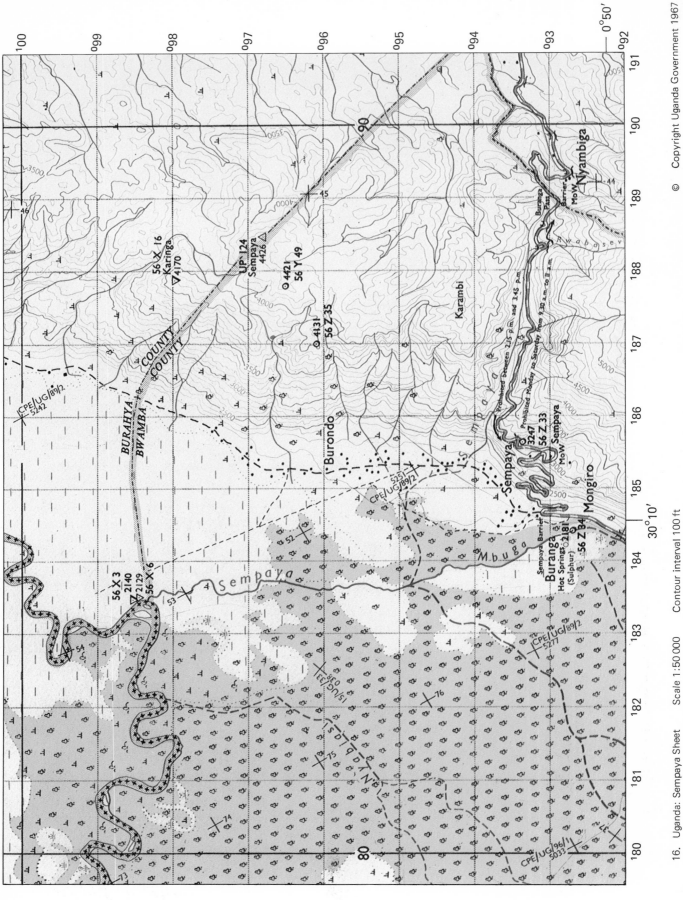

144

16. Uganda: Sempaya Sheet   Scale 1:50 000   Contour interval 100 ft

© Copyright Uganda Government 1967

4. Malawi: Zomba Sheet (cont.)

6. Describe the present dissection of the main highland masses by the drainage system. Draw two or three sketch-maps to show how the region would appear at future stages of dissection and write a description of each stage.

7. What is the origin of the Zomba Plateau to the north of Zomba? What reasons would you give to explain the fact that it is not a fault line scarp? Draw transects at several places along the scarp face to show the gradients and mark the free face and the debris slope.

5. Uganda: Kampala Sheet 1:50 000

Accordant summits

Turn to pages 28, 29, 68 and 122 and re-read the references to laterite landforms, planation in Uganda and the formation of Lake Victoria.

1. Lay a tracing paper over the map area and shade in the planation summits above 4250 ft and the last contour on hills lower than 4250 ft. Measure the distances between each of these summits. In the valleys, what is the height of the lowest contour line? Draw conclusions from these exercises.

2. Draw transects E–W running through Buziga (571 285) and Mutundwe (482 313). Comment on the accordant summit and valley-base levels. Then explain how these accordant levels were formed.

3. Measure the angles of the slopes on Mutundwe, Buziga, Makindiye (542 311) and Bunamwaya (497 286). Comment on your findings.

4. What is the cause of the Kansanga, Kyetinda and Kaliduubi channels? What type of coastline is this? Account for the offshore feature northeast of the headland at Kauko (600 315). What natural process is occurring along this coast?

5. Comment on the drainage pattern shown in the extract.

6. Zimbabwe: Pungwe Falls Sheet 1:63 360

River capture and nick points

Read pages 59, 64 and 65 and study Fig 123 before starting this exercise.

1. Locate the area in your atlas by referring to the latitude and longitude shown.

2. Draw a northwest to southeast transect through both the Pungwe valley above the Pungwe Falls and the Pungwe Gorge. Write a comparative description of the Pungwe valley and the Pungwe Gorge noting the differences in slope and depth of incision. Work out the average gradient along the two sections of the river above and below the Falls over similar distances. Relate your findings to the process of river capture.

3. To obtain a clear picture of the drainage pattern, draw a sketch-map by tracing off only the rivers in blue. Draw in the watersheds in a red pecked line and indicate directions of flow with small arrows. Mark in such features as: misfit stream, capturing stream, rejuvenated stream, elbow of capture, wind gap, etc.

4. Examine the map extract for future possibilities of capture by the Pungwe. Explain the possibility.

5. Compare the upper headwaters of the Nyakupinga River with the Pungwe River at the Causeway Cottages. Draw transects across both rivers. How much deeper is the Pungwe (the former upper Nyakupinga) than the present upper headwaters of the Nyakupinga? Relate this erosion difference to the process of river capture.

7. Kenya: Mount Kenya Sheet 1:25 000

Features of a glaciated upland

Read pages 72–9 before starting this exercise.

1. Locate this region in your atlas by referring to the latitude and longitude shown.

2. Calculate the surface area of the ice in the Lewis and Tyndall glaciers. Compare these with some of the major glaciers of Europe, North America and New Zealand. Using the blue contour lines draw sketches to show the glaciers in their cirques and valleys.

3. Draw a geomorphological sketch-map of the region indicating arêtes, tarns, pyramid peaks, cirques, glaciers, recessional moraines, glacial valley steps, tributary glacial valleys, headwalls, etc.

4. What evidence is there on the map that the glaciers were formerly much more extensive?

5. Measure the angles of slopes from the surface of the tarns to the arêtes on the cirques above them. Calculate the areas of the tarns. Draw cross-sections of each U-shaped valley, noting the changes in gradient and such features as shoulders.

6. Draw a three-dimensional block diagram of the Hidden Tarn, its cirque, and the part of the U-shaped valley shown.

7. What is the probable origin of the elongated lake east of the 37° 20′ longitude line on the Rupingazi River?

8. Attempt a map of the region recreating the extent of ice during the period of maximum glaciation (see Fig 135).

9. Imagine you are standing on Castle Hill looking northeastwards through an angle of 100°. Draw a sketch-map of the landscape you would be likely to see.

8. Tanzania: Arusha-Tengeru Sheets 1:50 000

A volcanic region

Read pages 98–103 before starting this exercise.

1. Locate this region in your atlas by referring to the latitude and longitude shown.

2. Draw an annotated transect along line 30. Work out the various gradients for the slopes of Kibwesi and Kiutu.

3. For each volcano on the map extract draw E–W cross-sections

for comparison of height, slopes, general shape. Comment on your drawings. Compare the heights of these volcanic mountains with hills and mountains in your own district.

4. Locate the following landforms on a sketch-map of the area:

a lava plateau; parasitic or subsidiary cones; craters; cones; crater lakes.

5. Imagine you are standing on Mama Doki Hill (508 322) looking south through an angle of 140°. Draw an annotated sketch of what you would probably see in the actual landscape.

6. What is the area of Lake Duluti? Draw a cross-section to show its morphological features. Explain how the feature has been formed. Why has not a similar but smaller crater lake formed in Kibwesi's crater?

7. Compare the valley of the Temi River with the others on the map and suggest reasons for the major differences.

9. Ghana: Wurupong Sheet 1:50 000

Drainage patterns and adjustment to structure

Read the section on pages 60–65 and 82. The map shows part of the Volta region of Ghana just to the north of Kpandu. It is a region where the drainage is adjusting to structure by vigorous river capture and incision.

1. Estimate the gradient of the escarpment overlooking the road from Kpandu between the spot height 471 ft just north of Wurupong and the triangulation point 2386 ft at Mlabo. Compare this with the gradient on the eastern side of the ridge descending to the Kpeyi River. Suggest possible reasons for the differences in steepness.

2. Draw transects through Wálibo in the north, through Malabo in the centre, and through Bamaleso in the south. Note the morphological features shown on your transects.

3. Draw a sketch-map of the drainage pattern indicating watersheds and direction flow.

4. What patterns are displayed by the drainage system? Can a pattern of subsequent and obsequent streams be worked out?

5. In what ways is the drainage system adopting itself to the structure of the region? Using one of the transects from question 2, indicate the possible rock structure of the region by noting the occurrence of hard or soft rocks and zones of rock weakness.

6. River capture has occurred and may occur in the future. This is indicative of variations in rock resistance. On your drainage sketch-map, identify by using symbols: a water gap, a wind gap, an elbow of capture, the capturing stream and the captured stream. There are two possibilities of future river capture. Identify these and justify your choice.

7. Analyse the basin of the Tsatsadu into first order, second order, etc., streams.

8. Draw long profiles for the rivers Kpeyi and Tsatsadu and compare the two.

9. Draw a transect E–W through Misrotaw indicating the position of river channels. Compare this with Fig 118. What do you infer from your transect?

10. South Africa: Oudtshoorn Sheet 1:500 000

Folded mountains and superimposed drainage

Note that the scale is very much smaller than the other maps in this section. Read pages 61–63, 83 and 84 before you start this exercise.

1. Locate the area of this map extract in your atlas by referring to the longitude and latitude shown.

2. Draw an annotated transect from north to south along the 22°E line of longitude. Insert the probable geological structure by referring to Fig 114.

3. Draw a sketch-map of the region to show the chief physical divisions. Number each of the divisions and describe each in turn. Try to explain the origin of these different landscapes in terms of their rock structure and geological history. Is there evidence of planation?

4. Draw a sketch-map to show the chief rivers and their major tributaries. Describe the patterns. How have the rivers
(a) adapted themselves to structure;
(b) not adapted themselves to structure?

11. Tanzania: Ngombezi Sheet 1:50 000

Dissected horst, river adjustment to structure, old river features

Read pages 61–3 and 88 before starting this exercise.

1. Locate this map extract in your atlas by referring to the latitude and longitude shown.

2. Compare this map with the Mazinde sheet on page 140. What similarities are there?

3. Draw a transect from the southwest corner to the northeast corner of the extract. Annotate your transect. Attempt an indication of the geology of the region.

4. Divide the region shown into its major relief zones and briefly describe each.

5. Draw a trace sketch-map of the drainage in blue and divide the drainage according to the nature of the streams and river patterns involved.

6. Describe the course of the Pangani River noting particular features along its course, measuring the amplitude of the meanders, etc. Why does the course of the Pangani's major northern tributary become discontinuous over long stretches?

7. The Pangani cuts through highland masses at two points. Suggest possible reasons for this.

12. Tanzania: Mazinde Sheet 1:50 000

Fault block, graben and inland delta

Read pages 50 and 88 before starting this exercise.

1. Locate the area of the map in your atlas by referring to the latitude and longitude shown.

2. Draw an annotated transect from the southwest corner of the map through spot height 3138 to the northeast corner of the map. Describe the relief shown and attempt to indicate the geology of the rocks in this region. Indicate the presence of faults.

3. Draw a relief sketch-map dividing the region into its three major natural relief sections. Describe the general relief of each of these regions and explain their origins.

4. Draw a trace sketch-map of the drainage pattern in blue. In what way has the drainage of the Mkomazi River been influenced by tectonic disturbance? Work out the gradient of the course of the Mkomazi from its point of entry to its point of exit on the map extract. Describe the drainage features of the Mkomazi and its tributary streams. Explain why the rivers from the highlands do not join the Mkomazi directly by visible channels. What type of soils would you expect along the footslopes of the fault scarps on the map? How would they be formed?

5. Draw a transect NW–SE along the line of the scarp face to illustrate how the rivers here are etching the face.

13. Kenya: Nairobi-Nakuru Sheets 1:250 000

Rift faulting and vulcanicity

Note the scale is much smaller than the other maps in this section. Read pages 88–93 and 98–103 before you start this exercise.

1. Locate the area in your atlas by referring to the latitude and longitude shown.

2. Draw two annotated transects:
(a) from east to west to pass through Mt Susua
(b) from north to south to pass through the Ngarowa Gorge and Mt Susua.
Insert on (a) the probable geological structure (refer to Fig 179 which was taken across a line 65 km to the south).

3. Write a description of the landforms due to volcanic action. Explain the origin of the 'raft' of solidified volcanic material in Susua's crater. Work out the total area of lava shown on the map extract in square kilometres. Draw a cross-section through mounts Orgaria and Longonot and compare their profiles. What is the significance of the hills immediately to the west of Mt Longonot?

4. Write a brief description of the landforms due to faulting. What type of faults are present? Work out the angle of slope of the fault scarps. Describe any evidence of faulting on the floor of the valley.

5. Discuss the drainage system of the map, using a sketch-map to show drainage only. What patterns emerge? What evidence is there of aridity in the region? Why is there so much intermittent drainage?

6. Imagine you are standing at trigonometrical point 6362 ft in the southwest corner of the map. You are looking northeastwards and your angle of vision is 50°. Draw a sketch of what you would see before you, and describe the scene.

14. Zimbabwe: Nkai Sheet 1:50 000

Pan formation and seepage in a dry savanna region

Read pages 65 and 107 before starting this exercise.

1. Locate the area of the map in your atlas by referring to the latitude and longitude shown.

2. Draw a trace sketch-map showing only the drainage pattern in blue. Draw red pecked lines to show watersheds. What is the general height of these watersheds?

3. What is the gradient of the long profile of the Nkai River from the tip of the first order stream just east of Maqaqeni School at point 915 872? Draw a long profile diagram. How does this gradient compare with rivers in your own area?

4. Analyse the drainage pattern of the upper Nkai basin into first order, second order, etc., streams. Work out the stream density of the Nkai basin as shown.

5. What is the geomorphological origin of the Isenka Ridge? Draw a transect from 860 856 southwest of Dankambiri Pan to 950 960 north of the Nkai's eastern exit from the map. Compare the ridge and its slope with similar features in your own area. On your transect, sketch in the probable lie of the water-table.

6. What effect have the man-made dams at Sikobakoba (932 880) and at Upper Nkai (915 915) had on the gradient of this stretch of the long river profile? Draw a sketch of the gradient to illustrate your answer.

7. What is the likely origin of the pans shown in the extract? (see p. 107). How are they related to the Nkai drainage? What would be their maximum areas? Compare these with the areas shown. What does the present drainage pattern and pan presence tell you about the surface and subsurface rocks and soils?

15. Kenya: Kilifi Sheet 1:50 000
Coral reefs and lagoons

Read pages 122–3 before starting this exercise.

1. Locate this region in your atlas by referring to the line of longitude and the line of the equator (00) shown.

2. Describe the general relief of the region and relate it to the geological section on p. 123.

3. Explain the formation of the Bandari ya Wali and the entrance of Kilifi Creek.

4. What is the origin of the sedimentary deposits in the Bandari ya Wali? State why these are probably of different origin than the sediments in square 97 00. Would there by any difference in the mineral contents of the coastal and interior beaches? Give reasons for your statements.

5. Measure the distance of the coral reef from the coast at its greatest length and from near the golf course. Explain the differences.

6. Obtain information on mangrove vegetation (Gresswell *Physical Geography*, p. 362, Harrison Church, *West Africa* p. 65 and 310). Then explain the effects of this vegetation on coastal aggradational processes.

16. Uganda: Sempaya Sheet
1:50 000

1. Locate the area of the map in your atlas by referring to the latitude and longitude shown.

2. Draw a sketch-map dividing the area into distinct physical areas. Then draw an annotated cross-section E–W along line 096. Can you guess what the geological structure is? Outline on your sketch-map an area where there is some degree of planation.

3. Study the highland zone in the east. Describe the general relief of this region noting evidence of planation. What is the gradient of the escarpment in the west of this area? Is this an erosion scarp or a fault-line scarp? Give reasons for your answer.

4. What tectonic landform is the highland mass? What evidence is there to show that geologically speaking it is a fairly recently formed feature? What type of rock is it likely to be?

5. Now study the lower western region. How do you account for the extreme flatness of this area? What type of rocks make up this area?

6. Draw a sketch-map of the region showing only the drainage patterns in blue. Describe the major river in the northwest of the map. Note any natural features which suggest it is an old river. What effect does the water-table have on this lower area?

7. Now study the highland drainage. Draw the watershed line on your sketch-map. Compare the valleys and gradients of those rivers flowing westwards with those flowing eastwards and account for any differences. Explain why the westward-flowing rivers end abruptly at the bottom of the escarpment. What is the reason for the upper Sempaya River cutting through the escarpment while the other rivers do not do so? What is the significance of the hot springs at Buranga?

Appendix: Classification of Rocks

TABLE 1 CLASSIFICATION OF IGNEOUS ROCKS

Acidic	Intermediate	Basic	Texture	Mode of Formation
A. EXTRUSIVE				
Volcanic Rocks Rhyolite: Grey, white, pink. Associated with sial[1] rocks of continents.	Andesite: Dark-coloured. Associated with sial rocks of continents.	Basalt: Heavy. Associated with sima[2] rocks of continental fringes.	Fine grained crystals less than 0·5 mm.	In lava flows.
B. INTRUSIVE				
(a) Hypabyssal Quartz Porphyry: Light coloured. Phenocrysts[3] of quartz, biotite, orthoclase, hornblende.	Porphyrites: Dark-coloured. Intrudes into andesite.	Dolerite: Dark-coloured Heavy.	Medium grained. Crystals between 0·5 and 1·25 mm.	In sills, dykes and laccoliths.
(b) Plutonic Granite: Very common. Light coloured. Over 90 per cent of phenocrysts are quartz and feldspar, some biotite.	Diorite: Greyish colour. Composed of plagioclase feldspar and ferro-magnesian minerals.	Gabbro: Has no quartz but plagioclase feldspar and pyroxene with olivine. Very dark, tough rock.	Coarse grained. Crystals larger than 1·25 mm.	In batholiths.

[1] The less dense rocks of continents whose main constituents are silicates and aluminium minerals
[2] The heavier, denser rocks of the lithosphere whose main constituents are silicates and magnesian minerals
[3] Large crystals

TABLE 2 CLASSIFICATION OF SEDIMENTARY ROCKS

A. MECHANICALLY FORMED ROCKS

1. Rudaceous
(pebbly to stoney, average grain size above 2 mm)

Breccia (angular rock fragments)
Conglomerate (rounded rock fragments)
Tillite (fossil boulder clay of glacial origin)

2. Arenaceous
(sandy, grain size 0·05–2 mm), rich in quartz grains)

Sandstones
Greywackes (quartz, feldspar grains and rock fragments in clay matrix)
Quartzite
Arkose (almost entirely feldspar grains)

3. Argillaceous
(muddy, grain size less than 0·05 mm)

Soft clays
Mudstone
Siltstone
Marl
Shales

B. CHEMICALLY FORMED ROCKS

1. Calcareous
Limestones (more than 50 $CaCO_3$)
Dolomites

2. Siliceous
Chert
Flint
Jasper

3. Ferric
Clay iron

4. Saline
Evaporites
Gypsum
Rock salt (halite)
Anhydrate
Potassium salts

C. ORGANICALLY FORMED ROCKS

1. Calcareous
Foraminiferous limestone
Coral limestone
Shelly limestone
Oolitic limestone (precipitated round sandgrain nuclei)

2. Carbonaceous
Lignite
Coal

3. Ferric
Iron ore
Ironstone

TABLE 3 CLASSIFICATION OF METAMORPHIC ROCKS

Original Rock	New Metamorphic Rock	Composition	Texture	Metamorphic Process
Shale	Slate Phyllite Schist	Numerous silicate minerals, e.g. serpentine, hornblende, micas, talc, chlorite. Minerals dark, flaky.	Very fine grained Fine grained Medium to coarse grained All foliated	Regional
Igneous rocks: greywacke, mica schist, arkose	Gneiss	Much feldspar plus quartz, and dark silicates—garnet, micas, pyroxenes, amphiboles.	Medium to coarse grained, foliated	Regional
Sandstone	Metaquartzite	Predominantly quartz	Medium to coarse	Regional or contact
Limestone, dolomite	Marble	Dolomite or calcite. Sometimes silicates of magnesium and calcium.	Medium to coarse	Regional or contact
Shale	Hornfels	Silicates of dark colour	Fine to very fine grained	Contact

Further Reading List

The following textbooks should form part of every geography library. They are intended not as supplementary readers but as extension readers, widening the student's knowledge of geomorphology in other continents. The student should attempt to read at least three or four of these books.

1. A. Holmes, *Principles of Physical Geology*, Thomas Nelson, 3rd edition, 1977
2. D. Q. Bowen (Editor), *A Concise Physical Geography*, Hulton Educational Publications, 1972
3. B. W. Sparks, *Geomorphology*, Longman, 1960
4. R. K. Gresswell, *Physical Geography*, Longman, 1967
5. G. Drury, *The Face of the Earth*, Penguin, 1966
6. A. N. Strahler, *Physical Geography*, John Wiley, New York, 1964
7. S. W. Woolridge and R. S. Morgan, *An Outline of Geomorphology*, Longman, 1959
8. P. Lake, *Physical Geography*, Cambridge University Press, 1971
9. N. K. Horrocks, *Physical Geography and Climatology*, Longman, 1962
10. I. G. Gass and others, *Understanding the Earth*, Open University Press, 1971

The chapters on geology and geomorphology in the following books on Africa:

11. R. J. Harrison Church, *West Africa*, Longman, 1966
12. W. T. W. Morgan, *East Africa*, Longman, 1973
13. A. T. Grove, *Africa South of the Sahara*, Oxford University Press, 1969
14. A. B. Mountjoy and C. Embleton, *Africa*, Hutchinson, 1966

In addition:

15. L. C. King, *South African Scenery*, 3rd edition, Oliver and Boyd, 1967
16. C. Willock, *Africa's Rift Valley*, Time Life Books, 1974

The student should also make reference to a good atlas when reading the text and Philips' *Modern College Atlas for Africa* is very suitable for this purpose.

Journals

The following articles are listed for the convenience of students wishing to read more deeply into the subject of Africa's geomorphology. Abbreviations used:

G.J. *Geographical Journal*
G.M. *Geographical Magazine*
I.B.G. *Transactions and Papers of the Institute of British Geographers*
J.T.G. *Journal of Tropical Geography*
Q.J.G.S. *Quarterly Journal of the Geological Society*
S.A.G.J. *South African Geographical Journal*
Z.G. *Zeitschrift fur Geomorphologie*

Weathering and Landforms

1. Laterite and Lateritic Soils, J. A. Prescott and R. L. Pendleton, Commonwealth Bureau of Soil Science, Technical Communication, 47, 1952, p. 35.
2. Vegetation and Geomorphology in Northern Rhodesia, M. M. Cole, *G.J.*, 1963, 290–310.
3. The Inselbergs of Uganda, C. D. Ollier, *Z.G.*, Vol 4, 1960, 43–52.
4. The Physiography of the Mengo District, J. W. Pallister, Buganda, *Uganda Journal*, 1957, 16–29.
5. Fringing Pediments and Marginal Depressions in the Inselberg Landscape of Nigeria, J. C. Pugh, *I.B.Q.*, 22, 1957, 15–32.
6. A Note on some Inselbergs around Bauchi, Northern Nigeria, M. O. Oyawoye, *Nigerian Geographical Journal*, 3(2), 1960, 33–7.
7. A Theory of Bornhardts, L. C. King, *G.J.*, 112, 1948, 83–7.
8. Down in the Forest (weathering in the humid tropics), M. F. Thomas, *G.M.*, Nov 1972, 135–9.
9. Savanna Lands between Desert and Forest, M. F. Thomas, *G.M.*, Dec 1971, 185–9.
10. Climate and the Geomorphic Cycle, A. A. Miller, *Geography*, 46, 1961, 185–97.
11. Landforms on the Sedimentary Rocks of Southwestern Nigeria, L. K. Jeje, *J.T.G.*, Vol 36, June 1973, 31–41.
12. Landform Development at the Boundary of Sedimentary and Crystalline Rocks in Southwestern Nigeria, L. K. Jeje, *J.T.G.*, Vol 34, June 1972, 25–33.
13. Transvaal Karst: Some Considerations of Development and Morphology, A. B. A. Brink and T. C. Partridge, *S.A.G.J.*, 48, 11–84.

Slope Retreat and Planation

1. Pediplanation and Isostacy, L. C. King, *Q.J.G.S.*, III, 1956, 370.
2. Isostatic Readjustment and the Theory of Pediplanation, J. C. Pugh, *Q.J.G.S.*, III, 1955, 361–74.
3. Slope Development in Uganda, J. W. Pallister, *G.J.*, 1956, 80–87.
4. Some Residual Slopes in the Great Fish River Basin, South Africa, G. Robinson, *G.J.*, 1966, 386–90.
5. High-level Surfaces in the Eastern Highlands of Nigeria, J. C. Pugh, *S.A.G.J.*, 36, 1954, 31–42.
6. Hill Slopes and Pediments of the Semi-arid Karroo, T. J. D. Fair, *S.A.G.J.*, 30, 1948, 71–9.

Rivers

1. Surface, Drainage and Tectonic Instability in part of Southern Uganda, J. C. Doornkamp and P. H. Temple, *G.J.*, 1966, 238–52.

2. The Nile and Murchison Falls, A. B. Ware, *G.J.*, 1967, 481-2.
3. Vicissitudes of the Course-Changing River, C. Embleton, *G.M.*, June 1972, 601-6.
4. Unfathomless Course of the Timeless Nile, I. Cranfield, *G.M.*, June 1972, 615-20.
5. The Kunene River and the Etosha Pan, J. H. Wellington, *S.A.G.J.*, 20, 1938, 21-33.

Lakes

1. The Lake Albert Basin, W. W. Bishop, *G.J.*, 1967, 469-80
2. Rise and Fall of Lake Chad, A. T. Grove, *G.M.*, 1969.

Glaciation

1. The Glaciers of Mount Baker, Ruwenzori, J. B. Whittow, *G.J.*, 1959, 370-79.
2. The Speke Glacier, Ruwenzori, J. B. Whittow and A. Shepherd, *Uganda Journal*, Vol 23, 1959, 153-61.
3. The Landforms of the Central Ruwenzori, East Africa, J. B. Whittow, *G.J.*, 1966, 32-42.

Tectonics

1. The Nyasa Rift Valley, F. Dixey, *S.A.G.J.*, Vol 23, April 1941, 21-45.
2. Planet of Cracks and Contortions, J. C. Doornkamp, *G.M.*, Oct 1971, 17-22.
3. Mountains of the Moon, R. A. Redfern, *G.M.*, 1962, 395.

Vulcanism

1. The Shira Plateau of Kilimanjaro, G. Salt, *G.J.*, June 1951, 150-66.
2. The Kivu Volcanoes in the Belgian Congo, T. W. Gevers, *S.A.G.J.*, Vol 22, 1940, 3-26.
3. Mount Cameroon: West Africa's Active Volcano, S. H. Bederman, *Nigerian Geographical Journal*, Vol 9, Dec 1966, No. 2, 115-27.
4. Safety Valves of the Earth, P. Francis, *G.M.*, March 1973, 438-42.

Deserts

1. The Ancient Erg of the Hausaland, A. T. Grove, *G.J.*, 1958, 528-33.
2. Geomorphology of the Tibesti Region, A. T. Grove, *G.J.*, 1960, 18-31.
3. Sand Formations in the Niger Valley between Niamey and Bourem, J. R. V. Prescott and H. P. White, *G.J.*, 1960, 200-203.
4. The Jebel Marra, Darfur, and Its Region, J. H. G. Lebon and V. C. Robertson, *G.J.*, 1961, 30-49.
5. Notes on the Geomorphology of the Northern Region, Somali Republic, J. W. Pallister, *G.J.*, 1963, 184-7.
6. Recent Expeditions to Libya by the Royal Military Academy, Sandhurst, M. A. J. Williams and D. N. Hall, *G.J.*, 1965, 482-501.
7. Across the Waterless Namib Dunes, R. A. C. Jenson, *G.M.*, 1970.
8. Uromantic Desert, Ronald Cooke, *G.M.*, 1971.
9. Conservation of a Desert Landscape, J. Gordon, *G.M.*, 1971.

10. Journey Across the Grand Erg Oriental, I. G. Wilson, *G.M.*, Jan 1971, 264-71.
11. Some Aspects of the Morphology of the Air Mountains, Southern Sahara, M. B. Thorp, *I.B.G.*, No. 47, Sept 1969, 25-43.
12. The Study of Desert Morphology, R. F. Peele, R. U. Cooke and A. Warren, *Geography*, No. 263, Vol 59, April 1974, 121-37.

Coasts

1. Quest in the Red Sea Coral Reefs, C. H. Roads, *G.M.*, 1969.
2. The Sandspits of Lake Albert, A. Shepherd, *Z.G.*, Vol 5, 1961, 53-72.
3. The Raised Beaches of the Peninsula of Sierra Leone, *I.B.G.*, Vol 31, 1962, 15-22.
4. Sand Movement in Relation to Wind Direction as Exemplified on the Nigerian Coastline, J. C. Pugh, Research Notes, Dept of Geog, University College, Ibadan. Vol 5, 1954.
5. A Classification of the Nigerian Coastline, J. C. Pugh, *J. of the W.A. Science Assoc.*, Vol 1, 1954, 1-12.
6. Coral Reefs: The Last Two Million Years, D. R. Stoddart, *Geography*.

Glossary

Note: This glossary only contains the most important words or those with which the student may be unfamiliar. Where a word has been adequately explained in the text, it is printed in the index.

ablation   Loss of ice from a glacier by surface melting and evaporation.

ablation moraine   Debris deposited by the above process.

abrasion   The wearing away of rock caused by moving debris, e.g. sand.

acidic lava   Lava with high amounts of quartz and feldspar; it solidifies fairly quickly to form volcanic cones.

aeolian   Related to wind, e.g. aeolian deposits have been transported and deposited by wind.

aerodynamic ripple   Miniature ridges and depressions caused on sand surfaces by unequal air pressure.

agglomerate   Masses of coarse, angular rock fragments caused by volcanic eruption.

alluvial fan   A half-cone shaped deposit of sediments with a low angle of rest. They have been deposited by a stream where it enters a plain from higher land.

alluvium   Sand and silt deposited by a river.

alveoles   Small depressions pitting a rock surface caused by chemical weathering and sand blasting.

amygdale   A concentration of fossilised chemicals in the space formerly occupied by air or gas bubbles in lava.

antecedent drainage   A river system which has maintained its flow across a land surface that has been raised by folding, faulting or upwarping.

anticline   An arch-like upfold in rock strata.

anticlinorium   A giant arch of rock strata whose limbs are wrinkled into numerous smaller folds.

arcuate (or cuspate) delta   A fan-shaped delta whose coast forms an arc.

arête   A jagged, knife-edged ridge usually lying between two cirques.

arête and pinnacle karst   A landscape of limestone beds composed of sharp, jagged ridges separated by broad flat regions of erosion.

asthenosphere   A relatively fluid layer of the earth's mantle lying below the lithosphere; movement in the asthenosphere caused by convection may cause plate movement.

backshore bar   A ridge of sand, formerly a sand bar, which has migrated shorewards and now encloses coastal marshes.

badlands   Arid plateau deeply dissected by gullies and caused by occasional rainfall.

bahada   Spanish word for a piedmont alluvial plain formed by the merging of several alluvial fans.

barrier beach   A sand or shingle bar exposed even at high tide.

basal sapping   Gradual undercutting at the base of a rock mass by weathering and debris removal by water.

base-level   The lowest level to which a river can erode its bed, usually sea-level.

basic lava   Lava which contains much mica, hornblende, augite and olivine. The lava is more fluid than acid types.

batholith   A large dome-shaped mass of magma which accumulates beneath the surface and acts as a reservoir or chamber supplying magma and lava to volcanoes, lava flows, etc.

bauxite   The ore of aluminium, a type of laterite rich in aluminium compounds; chemical formula $Al_2O_3\ 2H_2O$.

bedform   Landforms created in sand by wind currents.

bed load   The rocks, pebbles, sand, etc. moved along the bed of a river.

bergschrund   A large crevasse at the head of a glacier where the ice moving down the valley pulls away from cirque walls.

berm   A ridge of coarse sand and stones built up at the upper swash line by strong waves.

bevel   An erosion level or plane of erosion.

biological weathering   Weathering caused by plants, animals and bacteria.

blind valley   A valley in a karst region, with or without a river, which ends in a sinkhole.

bornhardt   A domed inselberg forming a prominent mass of crystalline rock standing above a pediplain.

boss   A small batholith or a small projection from a batholith. General term for a resistant mass of rock.

bottomset beds   The fine, thin layers of sediment in each layer of a delta.

boulder clay   Rocks, stones, pebbles and small boulders embedded in finer clays left by ice when it melts.

braided channels   A number of narrow, interconnecting channels formed where the river begins to deposit silt to form sandbanks.

breaker   A wave breaking into foam as its rotary motion is broken in shallow water.

breccia   Italian word for angular fragments of rock embedded in finer material.

butte temoine   French word for the remnant conical hills of the last stage in the erosion of karst landscapes.

calcrete   A type of laterite rich in calcium, found in dry areas.

caldera   A circular, steep-walled basin, usually several kilometres across, with steep infacing scarps, caused by subsidence of the upper part of a volcano.

capillary action   The migration of minute drops of water through the voids (air spaces) between rock or soil grains. This may be upwards from the saturated rock into a zone called the capillary fringe.

carbonation   The process whereby salts are formed by the mixture of carbonic acid with a base.

case hardening   The formation of a thin, hard layer of calcium carbonate over the surface of limestone forms in caverns.

castle or castellated koppie   A hill or ridge of large, angular boulders caused by the weathering of jointed crystalline rock.

centripetal drainage   A stream network which flows towards a central point, e.g. towards the centre of a basin.

chott or shott   Arabic word for a shallow, temporary salt lake or marsh found on the plateaus of Algeria and Morocco and to the south of the Atlas Mountains.

cirque   French word for deep, rounded, bowl-like hollow on the side of mountains, formed by erosion by snow or ice.

clastic   Word describing fine debris particles derived from weathered rock or weathered skeletal remains of organisms.

clint   A ridge of bare limestone rock, usually a metre or so wide, between fissures or grikes.

composite moraine   A glacial moraine formed from several other types of moraine, e.g. where lateral and median moraines join on a glacier surface.

compound sinkhole   An enlarged sinkhole in karst areas formed by the merging of several smaller sinkholes.

concordant   Adjective describing landforms or drainage systems which are in alignment with rock structures, e.g. a concordant coastline is roughly parallel with the mountain ranges nearby, concordant drainage flows with the structure instead of across it.

conduit   A low-roofed, narrow passageway, usual formed by solution of the limestone along a bedding plane, linking one cavern with another.

cone karst   Sometimes called kegel karst (German). A karst landscape in which the remnant hills are cone-shaped.

consequent stream   A stream or river which follows the initial slope of the land or the dip; the river is a consequence of the direction of slope.

constant slope   The slope below the scarp or free face of a rock mass which is strewn with fallen debris.

contact metamorphism   A change occurring in rocks caused by the heat of an intrusion or injection of igneous rock.

core-stone   A stone, usually round in shape, formed from the chemical weathering below ground of a block of rock.

corrie   North English word for cirque.

corrosion   Chemical weathering of its channel by a river.

crag and tail   A hill with a steep face caused by plucking by a glacier and a gentle slope of the opposite side caused by moraine deposition in the lee of the rock.

craton   From the Greek word for stability; a large region of a continent which has remained undisturbed for over 150 million years.

creep   The imperceptible movement of the regolith down a slope.

crevasse   French word for a deep, wide crack in a glacier surface.

crevice karst   A limestone surface deeply dissected along joints to form narrow ridges and deep narrow gulleys.

cuesta   Spanish word for a ridge with a gentle dip slope on one side and a steep scarp face on the other. It is formed by the erosion of gently dipping strata in which the harder rocks remain as uplands or by the erosion of one limb.

cuspate delta   See arcuate delta.

cwm   Welsh word for a cirque.

debris   French word for detritus — rock fragments removed from rock surfaces by disintegration.

deflation   The lifting up and transport of sand and dust by wind.

deflation hollow   A shallow depression on a sand or rock surface due to removal of loose materials by wind.

dendritic pattern   A river pattern which resembles the branching roots of a plant or the branches of a tree.

denudation   The wearing away of land surfaces by running water, wind, moving ice, the sea, rain, frost.

detritus   Latin-derived word for debris.

differential erosion   The wearing away of rock surfaces at different rates because of varying resistance of the rock or the concentration of the eroding forces on particular parts of the rock face.

discharge   The amount of water in a river which passes a given point in a given unit of time.

discordant   The opposite of concordant; in which the rivers or coastlines run in a direction across the trend of valleys and ranges.

distributary   A branch of a river which leaves the main course and does not rejoin it, e.g., on a delta or an alluvial fan.

doline   A round hollow in a karst region caused by solution of the limestone at or near the surface.

domed inselberg   See bornhardt.

donga   A steep-sided dry water course or gully; similar to a wadi.

downwarp   An extensive depression in the earth's crust caused by gentle lateral pressures.

draa   A giant sand dune; the largest desert bedform.

dreikanter   German word for a three-sided ventifact — a pebble or stone worn by wind-blown sands so that it develops flat faces.

drumlin   A low, elongated hill of boulder clay, possibly formed in hollows in ice sheets.

duricrust   A layer of concentrated rock minerals cemented into a hard band by oxydisation.

dwala   African word for a whaleback — a long, low, rounded inselberg of crystalline rock.

Dwyka tillite   Fossilised moraine deposits with a maximum thickness of about 800m at the base of the Karoo rock system. It is widespread in southern Africa.

echo dune   A dune which forms in the comparatively calm air behind an object.

effusive eruption   A volcanic emission of lava which slowly wells out of the vent and is not accompanied by explosions.

ejectementa   Latin word for cinders, solidified lava and other rock pieces flung into the air by a volcanic eruption.

elbow of capture   The sharp bend formed where a capturing stream draws off the headwaters of a captured stream.

eluviation   The washing downwards by percolation of material in solution or suspension through soil layers.

englacial moraine   Stones, pebbles, small boulders, and sand transported within a glacier or ice sheet.

erg   Arabic word for a sand sea: a vast area covered by sand dunes.

erosion scarp   See cuesta.

esker   A long line of sands and gravels forming a low ridge. It was once the bed of a river flowing beneath an ice sheet.

estuarine delta   A delta which forms within the estuary of a river and composed of sand banks divided by narrow channels.

etchplain   A plain of basement rocks in the form of inselbergs, ruwares, koppies exposed by the stripping of weathered materials.

eustatic change (of sea-level)   Changes caused in the level of the sea by increases or decreases in the amount of sea water due to freezing and melting in the glacial and interglacial periods.

evaporites   Sedimentary rocks formed from the evaporation of lakes or seas saturated with minerals, e.g. rock salt.

exfoliation   The weathering of rock by the peeling off of thin layers of rock by heating, cooling and water action. From the Latin word *folia*, a leaf.

extrusive volcanic forms   Landforms caused by the solidification of lava above the earth's surface.

eye   Word used in South Africa for a spring or rise of water in a karst region.

fault breccia   Rock fragments formed by the disintegration of rock between two parallel, closely spaced faults.

fault-line scarp   A scarp produced by the erosion of less resistant rocks on one side of a fault-line after the original fault's escarpment has been worn away.

fault splinter   A partial exposure of rock strata by the sagging of the surface rocks on one side of a fault-line.

ferricrete   A laterite band rich in iron compounds.

fetch   The length of a stretch of surface water across which a wind is blowing.

firn   German word for névé — granular snow and ice accumulated at the head of a glacier valley which is transformed into glacier ice by pressure.

flash flood   A sudden flood of water, often unexpected. It occurs in arid and semi-arid regions when water from a sudden rain storm is focussed in a wadi or donga.

flow moraine   Lighter debris carried across the surface of a glacier or from its edges by flowing melt-water.

fluvio-glacial deposits   Debris washed out by melt-water from the snout of a glacier or edge of an ice sheet.

foliation   The layer structure of metamorphic rock, the layers being thin and usually parallel.

foot cave   Cave at the base of karst towers, possibly formed by water draining through the tower to reach the water table then flowing out from the base of the tower.

foreset beds   The coarse, steeply sloping, thick layers of sediments in a delta.

foreshore bar   A ridge of sediment or shingle accumulation roughly parallel to the shoreline with its surface above water level.

fossil dune   Ancient sand dunes formed during a drier period and now lying in a moister region. They may have been colonised by vegetation and become fixed.

free face   The almost vertical scarp face of the edge of a retreating mountain mass from which rock is being eroded.

frost riving   The disintegration of rock by the expansion of water in cracks when frozen.

full   Low upswells of sandy beach aligned at right angles to the wave approach angle.

fumarole   A vent or hole in the ground surface from which gas and steam are emitted under pressure.

Galleried caverns   Long narrow caves formed along bedding planes or at the water-table level in limestone rocks.

gargaf   Libyan word for broken, dissected blocky landscape usually found in limestone areas.

gibber plain   Australian word for a stony surface in arid regions.

glacial drift   Sediments deposited by glaciers and ice sheets or by water (lakes, rivers, streams) of glacial origin.

glacier mills   A vertical channel in a glacier enlarged by surface streams carrying debris. It usually develops from a crevasse.

Gondwanaland   The name given by the Austrian, Edward Suess, to the united continent of Africa, India, South America, Australia and the East Indies. Named after a region of India.

gour   Mushroom-like rock forms in arid regions; formed by erosion of lower rock section near ground by sand blasting.

graben   German word for downthrust fault block lying between horsts.

granular disintegration   The breakdown of rock under weathering into individual grains, each grain being prised loose from the rock surface.

grid faulting   Faulting occurring in a rectangular pattern, the fault lines crossing each other at angles.

grike   A fissure between clints on a limestone surface; caused by rainwater dissolving the rock along a joint.

ground moraine   Stones, small boulders, sand, etc. lodged in the sole or base of a glacier; also called lodgement till.

hammada   A desert area consisting mainly of bare rock which has been cleared of sand by the wind.

hanging valley   A tributary valley which enters a main valley side a considerable height above the main valley floor. Chiefly in glaciated regions and caused by the tributary glacier being unable to erode as deeply as the main valley glacier.

hard pan   A hard, impervious layer of soil, cemented by water and chemical action, lying some distance below the surface.

haystack hill   Remnant conical hill of last stage of erosion cycle in a karst landscape (see butte temoine).

head of rejuvenation   The point along a river's course or long profile which a young, vigorous river has reached by eroding backwards; the younger river has been rejuvenated (made younger) by an increase in the land surface's elevation above base-level.

head of water (on a river)   The vertical distance which the water in a river drops from source to mouth.

headcut   The low escarpment caused by backward erosion at the source of a river.

headward erosion   The process whereby a river lengthens its course by eroding backwards at its source.

helictite   An irregular-shaped calcium carbonate accumulation hanging from a cavern roof caused by the irregular accumulation of calcium carbonate particles.

hollow lake   Small lake which forms in the overdeepened section of a cirque; also called a tarn.

honeycomb weathering   The pitting of a rock surface with numerous small holes or alveoles by chemical weathering.

horn   German word for a pyramid peak in which adjacent cirques cut back into a mountain, each cirque separated by a sharp ridge or arete.

hum   Yugoslavian word for a haystack hill.

humic acid   A weak acid derived from the action of rainwater on decomposed vegetative and animal matter in the soil.

hydraulic action   The action of water pressure in weathering processes.

hydrolysis   Chemical decomposition which takes place when a chemical compound changes into other compounds by taking up water. The chemical formula for hydrolysis is $H_2O = H+ + OH^1$.

igneous intrusion   A penetration of molten rock into overlying strata.

impact ripple   Small wave shapes formed on a sand surface by the bombardment of sand grains.

impermeable rock   A rock which, because of the closeness of its grains or minerals, will not allow water to pass through. A rock may be impermeable but may still let water through joints and fissures; it is then pervious.

incised meander   A meander which has become deepened by uplift of the land and the rejuvenation of the river.

initial surface   The original surface of the landscape before erosion by rivers began.

inland delta   A large area of sedimentary deposition with braided channels in the middle course of a river; the sediments may be derived from an old lake bed or be due to deposition caused by a lessening of the river's gradient by back-tilting of the land.

inselberg   German for 'island mountain' — an isolated hill or group of hills standing above a level plain.

insolation   Energy received from the sun by the earth's surface.

interpluvial period   A long dry period of geological time occurring between two rainy or pluvial periods.

inversion relief   Relief in which land which was formerly at a high altitude has been so eroded that it is now lower than land which was formerly higher, e.g. an eroded anticline.

isostatic uplift   The gradual rise of a continental mass caused by the removal of eroded material and its deposition on the continental shelf; the weight of the deposits causes lateral pressures through the underlying softer parts of the earth's crust which cause a steady rise of the continent.

joint   A crack in a rock mass which occurs along a plane of weakness or joint plane and where no lateral or vertical movement has occurred to cause a fault.

kaolinite   A very fine whitish clay caused by the decomposition of feldspars in granite.

Karoo System   A rock system formed from sediments deposited from Middle Carboniferous times to the Triassic, with some volcanic intrusions. It covers about half the area of South Africa. The word 'Karoo' is also used for the Great and Little Karoo basins in South Africa.

karst   A limestone landscape in which there is virtually no surface water, most of the drainage being underground. The surface is deeply dissected or pitted with swallow holes, karst windows, poljes, etc.

kegel karst   German word for cone karst in which erosion has produced cone-like hills in a karst landscape.

koppie or kopje   Afrikaans word for an inselberg.

laccolith   A mushroom-shaped mound of intrusive magma below the earth's surface; the surface strata may be arched up in a dome.

lacustrine deposit   A deposit composed of lake sediments.

lagoon   Narrow strip of water separated from the sea by a reef or sand-bar.

lapies   French word for the grooves and ridges on a limestone surface; see clint and grike.

lapilli   Small rock fragments, varying in size from a pea to a small hen's egg, ejected from a volcano.

lateral moraine   Rocks, stones and other debris carried along at the surface and at the sides of a glacier.

laterite   A reddish-coloured hardened concentration of weathered rock minerals containing hydrated oxides of iron and aluminium. It occurs near the ground surface in thick layers and is soft when exposed but hardens under atmospheric influence.

leaching   The washing down of soluble substances such as mineral and organic salts, from the upper to the lower layers of the soil.

lee dune   See echo dune.

limb   The side of an anticline.

lithosphere   The solid crust of the earth, consisting of the soil and the mass of hard rock beneath, which acts as a thin skin to the more plastic and molten rocks of the earth's interior.

long profile (of a river)   A line which follows the bed of a river from source to mouth.

longshore drift   The migration of sand and shingle along a beach; the movement is caused by oblique waves which carry the material at an angle up the beach, but return it due to gravity at right angles to the shore line.

lunette dune   A semi-circular dune of fine sand which forms downwind from a pan.

magma   The molten rock which lies below the solid crust of the earth.

medial moraine   Debris carried along on the surface of a glacier and occupying a central position. It is formed by the joining of lateral moraines of two uniting glaciers.

mesa   A Spanish word for a flat, table-like upland, usually with with concave slopes on all sides, whose upper rock strata have resisted erosion.

microfissures   Tiny surface cracks in rock caused by weathering processes.

misfit stream   A stream or river which is too small for the valley in which it flows: its headwaters may have been drained off by a more vigorous stream.

mogote   Spanish word for a haystack hill.

monadnock   Red Indian word used by W.M. Davis to describe an isolated hill standing above the surrounding eroded countryside (an inselberg). Named after Mount Monadnock, New Hampshire, U.S.A.

monocline   A fold in rock strata in which the beds dip steeply then level out to resume their former incline.

moraine   The debris carried and deposited by ice sheets and glaciers.

morphology   The surface shapes of the landscape; the undulations of the earth's surface caused by various landforms.

moulin   French for a glacier mill.

nebka   Small dune formed where sand piles against a boulder or bush.

nested cone   Volcanic cone lying within a larger volcanic crater.

névé   French word for firn.

nick point   See 'head of rejuvenation'.

nivation   Erosion caused by snow.

nivation cirque   A cirque caused by the alternate freezing and thawing around a patch of snow in a hollow.

nivo-karst landforms   Landforms similar to those formed in karst regions but caused by the freezing and melting of snow in limestone regions.

nunatak   An isolated mountain peak projecting above the surface of the snow-covered ice at the edge of an ice sheet.

obsequent stream   A tributary to a subsequent river which flows in an opposite direction to consequent streams.

offshore bar   A long ridge of sand submerged beneath the sea and near to the coast.

orogen   An extensive region of rock strata which has remained stable for a considerable time, usually more than 1,500 million years.

orogenesis   The process of mountain building either by folding or by block faulting.

orogenic belt   A range or ranges of mountains in an extensive linear zone formed by orogenesis.

orogeny   A period during which considerable orogenesis took place.

overburden   The rock strata or consolidated debris overlying lower strata.

pan   Circular depressions in arid regions, usually on flat surfaces.

Pangaea   The name given by Wegener to the combined continents of the northern and southern hemisphere.

pediment   A gently inclined plain which lies at the foot of mountain escarpments; it is most common in arid and semi-arid regions.

pediplain   An extensive almost level plain caused by the retreat of escarpments and the extension of pediments.

peneplain   Meaning 'almost a plain'; a truly flat plain can never be attained, slopes being essential for water flow and waste removal.

periglacial   Describing climate or landforms occurring on the fringes of true glacial regions.

permeable rocks   Porous rocks which allow water to soak into them.

piedmont alluvial plain   An extensive inclined plain at the foot of an escarpment caused by the merging of several alluvial fans.

pipe   A narrow, tube-like vent filled with solidified lava.

playa   Spanish word for a pan or chott.

plunge pool   Deep pool caused by scouring of the river bed beneath a water fall.

plutonic rock   Igneous rocks which have solidified at great depth in the earth's crust; slow cooling produces a coarse-grained crystalline rock, e.g. granite.

pluvial period   A long rainy period during which the action of flowing water has a permanent effect on landform development.

polje   A long, flat-bottomed hollow in a karst surface which occasionally fills with water to form a karst lake.

pot-hole   Hole eroded in the solid rock of a river bed by the abrasive action of stones being swirled around by the current.

pressure release sheeting   The breaking away of sheets of rock from the rock face along joint planes caused by the rock expanding slightly. The expansion is due to the reduction of the pressure formerly caused by overlying rocks which have been eroded.

pyramid peak   See horn.

pyroclasts   See ejectementa.

quarrying or plucking   The removal of loose fragments or rock from the rock surface by the passage of ice.

radial drainage   Drainage pattern produced by rivers flowing outwards in all directions from a central upland.

raised beach   A beach which has been uplifted above its former level by earth movements, e.g. by isostatic uplift, to form low cliffs and a narrow coastal plain.

recessional moraine   The glacial debris left along the floor of a glacial valley by a receding glacier.

recumbent fold   An overfold in which the axial plane of the fold is almost horizontal and the fold limbs are almost parallel.

reg   Arabic word for the flat, extensive level plains from which finer material has been removed by wind to leave a carpet of stones and rocks.

regional metamorphism   Change occurring in rocks caused by shearing and pressures during mountain building over a wide region.

regolith   The zone of loose, non-cemented mineral grains and rock particles, the product of sub-surface weathering, which overlies the bedrock.

rejuvenation   The renewal of erosive activity, especially by rivers. Uplift increasing the head of water or an increase in discharge caused by stream capture will cause rejuvenation of a river.

rhourd   Arabic for a giant sand mountain.

ria   A long, narrow inlet on the coast caused by the sub-mergence of the land or a rise in sea level so that valleys are drowned.

ring dyke   A semi-circular, vertical or near vertical intrusion of lava in the rock beds, usually around a volcanic plug.

rip current   A strong, concentrated backwash from a beach which flows underneath the approaching waves.

rise   See eye.

river profile or gradient   The slope of the bed of a river from source to mouth.

roche moutonnée   French word for a low mound of rock in a glacial valley or on a glaciated plain which has been smoothed by the glacier on the approach side and plucked on the lee side.

rock flour   Very finely ground debris produced by the abrasion of rock surfaces by an ice sheet or glacier.

roller   Very large waves formed out at sea with up to 15m between crest and trough.

ruware   African word for a relatively smooth, slightly convex rock surface which rises a few metres above ground level. This may later become the summit of an emerging born-hardt.

salina or salt pan   A shallow hollow in a semi-arid region which formerly contained saline water but now has a deposit of salt caused by evaporation.

saltation   The bouncing motion of sand grains as they move over the desert surface.

sand sea   See erg.

sandspit   A long sand bar extending from a promontory on the coast, parallel to the coast and caused by longshore drift.

scoriae   Cinder-like rocks formed from lava, which contain numerous holes caused by gas and air bubbles which formed before the lava completely solidified.

seif dune   A long, sharply-ridged dune lying roughly parallel to the prevailing wind direction.

selective deflation   The removal of lighter particles of sand and dust by the wind which leaves heavier, denser particles behind.

sheet erosion   Erosion of land surface caused by overland flow of water in broad sheets.

sial   The comparatively light rocks of the lithosphere; from silicon and aluminium, the main elements of such rocks.

silcrete   A laterite with a high silica content.

sima   The heavier rocks of the lithosphere, e.g. basalt, lying beneath the sial and the ocean basins; from silicon and magnesium, the two major elements in these rocks.

sink hole   Saucer-shaped hollow in a limestone region in which water collects and enters underground caverns.

slickenside   The rock surface smoothed and scratched by friction between moving blocks of rock in a fault.

slip-off slope   The gentle deposition slope rising from the inner side of a meander.

snow-line   The lower limit line on highlands below which the temperature is sufficiently high to melt snow.

solfatara   The vent of a volcano which has ceased to emit lava but gives off gases and steam.

solifluction   The flow of soil, particularly in cold, wet climates where the soil remains saturated in its upper layers, and freeze-thaw takes place.

stack   A rock pillar or rocky island just offshore, isolated from the main body of rocks by wave erosion.

stalactite   An icicle-like column of accumulated calcium carbonate particles hanging from the roof of a limestone cavern.

stalagmite   A column or mound of calcium carbonate on the floor of a limestone cavern.

step faults   A series of normal faults which have formed parallel to one another, their surfaces forming a series of steps on the valley side.

stock   See boss.

stony desert   See reg or gibber plain.

stoping   The absorption of large blocks of surrounding rocks into an igneous intrusion by melting.

strand line   The line of an ancient beach or shore.

stria or striation   Marks scratched on rock surfaces by stones and other moraine embedded in a glacier or ice sheet.

subsequent stream   A tributary to a consequent river flowing at an angle to the consequent river and usually occupying a line of weakness in the rock.

sudd   A floating mass of vegetative matter found on the White Nile which breaks up into moving islands.

superimposed drainage   A drainage system which developed on a landsurface now removed by erosion, and which flows on the exposed underlying rocks.

supraglacial moraine   General term for all the moraine carried on top of a glacier or ice sheet.

swash   The movement of sea water up the beach after the wave has broken.

syncline   The trough of a fold in rock strata.

synclinorium   A huge trough like a syncline in which the limbs have been wrinkled into smaller folds.

tafoni   Indentations caused by chemical weathering on a rock surface; they are larger than alveoles and are somewhat like miniature caves.

tamarisk mound   See nebka.

tarn   See hollow lake.

tear fault   A fault in which the movement is horizontal along the plane of the fault.

tectonic arch   A huge upwarping of the earth's crust, several hundreds of kilometres across.

tectonic processes   Processes which build up features of the earth's crust — warping, fracturing, faulting.

terra rossa or **roxa**   A reddish soil with a high humus content.

thrust fault or **reversed fault**   A fault in which the upper rock beds have been pushed forward some distance over the lower strata.

till   See boulder clay.

tillite   A fossilised till or boulder clay several million years old and widespread in southern Africa due to the deposition of the Dwyka ice sheets.

tombolo   An island joined to the mainland by a sand-bar.

tower karst or **turm karst**   A karst landscape which has been eroded into tower-like pinnacles of remnant limestone.

traction   The movement of the bed load of a river along the bed of the river.

trapdoor fault or **hinge fault**   A fault only partially formed, the downthrown block surface resting at an angle to the horizontal.

travertine   The deposits of calcium carbonate in a cavern which form variously shaped mounds.

trellised drainage pattern   A river pattern which is rectangular in form, the bends and junction points forming sharp angles.

truncated spur   The spur of a former river valley which was planed off by a glacier later moving down the valley.

U-dune or **parabolic dune**   A ridge of sand in which part of the ridge has been blown away; the result resembles a crescent-shaped barchan but the arms point upwind.

uvula   A large depression in the surface of a karst region somewhat smaller than a polje.

vent   The narrow pipe of a volcano leading to the crater from below.

ventifact   A stone or pebble which has been shaped by the wind and which has several flat abraded faces.

volcanic bomb   A piece of molten lava which cools and solidifies as it twists through the air.

volcanic neck or **plug**   A mass of solid lava which blocks the main vent of a volcano.

wadi   A surface channel in an arid or semi-arid region which occasionally carries water after a storm.

waning slope   See pediment.

warp   An uplift of the earth's crust, convex in form, caused by gentle lateral pressures.

water gap   A narrow valley cut through a ridge.

water table   The variable and uneven surface in rock below which the rock is saturated with water.

wave-built platform   A flat broad surface formed by the build-up of debris deposited by sea action at the edge of a wave-cut platform.

wave-cut platform   A flat, broad rock surface left by the retreat of cliffs under sea attack.

waxing slope   The crest of a rock mass lying above the scarp face.

weathering front or **basal front**   The sub-surface zone at the base of the regolith where chemical weathering is attacking unweathered rock.

whaleback   See dwala.

wind gap   An abandoned part of a stream's course usually lying across a ridge. The stream which occupied the gap may have been captured.

xenolith   A large block of rock which has been partially absorbed into an igneous intrusion by stoping and which has not been fully melted.

yardangs   Ridges of rock with overhanging upper surfaces, grooves or deep corridors, and irregular-shaped rock forms sculpted by wind-borne sand. They lie roughly parallel to the wind direction.

zeugen   German word for pedestals of rock supporting table-like slabs of more resistant rock. The pedestals are of less resistant rock and have been worn by wind-borne sand.

zibber   Long, low (a few centimetres high) linear forms on the desert surface elongated in the direction of the wind.

Index

Significant references only are listed. Page numbers in italics refer to a diagram or photograph.